FOOD AND BEVERAGE CONSUMPTION AND HEALTH

SEED OIL

BIOLOGICAL PROPERTIES, HEALTH BENEFITS AND COMMERCIAL APPLICATIONS

FOOD AND BEVERAGE CONSUMPTION AND HEALTH

Additional books in this series can be found on Nova's website under the Series tab.

Additional e-books in this series can be found on Nova's website under the e-book tab.

FOOD AND BEVERAGE CONSUMPTION AND HEALTH

SEED OIL

BIOLOGICAL PROPERTIES, HEALTH BENEFITS AND COMMERCIAL APPLICATIONS

ALEXIS VARNHAM
EDITOR

nova publishers
New York

Library of Congress Cataloging-in-Publication Data

ISBN: 978-1-63463-056-6

Library of Congress Control Number: 2014950597

Published by Nova Science Publishers, Inc. † New York

Contents

PREFACE

The importance of fats for humans, animals and plants lies in their high content of energy. In addition, fats allow humans and animals to consume fat-soluble vitamins and provide them with essential fatty acids (FAs), which are indispensable because their bodies are unable to synthesize themselves. Vegetable oils are used for many food and industrial purposes. Although a wide variety of sources of vegetable oils, global consumption is dominated by palm, soybean, rapeseed and sunflower oils. In recent years there has been development of underexploited promising plant species as a source of dietary or specialty oils. Many of them contain significant quantities of oils and/or a high proportion of nutritionally, medicinally or industrially desirable FAs. This book discusses the biological properties, health benefits and commercial applications on seed oils.

Chapter 1 – The vegetable oils are dietary sources of sterols, vitamin E, and unsaturated fatty acids. The fatty acid composition and the content of an unsaponifiable substance in oilseeds depend on the variety of plant, degree of ripening seeds and the climatic conditions. Vegetable soybean *(Glycine max* (L.) Merr.) is an annual dicotyledonous plant belonging to the family of Fabaceae (Leguminosae) in the genus; their genotypes are divided into two categories: large-seeded and small-seeded. The large-seeded types are used for the fresh market in urban areas, whereas small-seeded types are used to make soybean sprouts. There are several groups of nutrients in soybean that are currently under investigation for healthy benefits, such as flavonoids and isoflavonoids, phenolic acids, phytoalexins, phytosterols, proteins and peptides, and saponins. In addition, soybeans are also an important source of minerals copper, manganese, molybdenum, phosphorus, potassium, B vitamin, and omega-3 fatty acids (alpha-linolenic acid). Replacing meat and dairy with soy may decrease total cholesterol intake by about 123 mg/day and saturated fat by about 2.4 g/day. These changes would reduce the risk of developing cardiovascular diseases, and even the cancer. Soybean oil is the edible oil extracted from soybean seeds, largely used as cooking oil in the world according to the US agricultural services. Dry soybean seeds compose 18-20% of extractable oil by weight. The seeds are then subject to pressing to obtain oil and the residue is used as animal feed. The crude soybean oil is yellow in color and contains moisture, lecithin, free-fatty acids, and some volatile compounds. These impurities have to be removed to obtain acceptable standard oil. Soybean oil has a good lipid profile with saturated, monounsaturated and polyunsaturated fats in healthy proportions (SFA:MUFA:PUFA=16:24:58). Linoleic acid (omega-6) is the major polyunsaturated fatty acid found in oil; phytosterols, especially B-sitosterol, inhibit cholesterol absorption and reduce blood LDL-cholesterol levels by 10% to 15%; anti-oxidant vitamin E, a powerful lipid soluble vitamin, is important to maintain the integrity of cell membranes and protect them from harmful reactive oxygen-free radicals; vitamin K, an essential element in promoting bone formation and strengthening, and neuronal protection in the brain.

More recently, genetic engineering is becoming an important tool to produce different and exotic oils derived from a diversity of plants in domesticated and commercial seeds-like soybean, by using genetic transformation. This technique has attracted commercial interest and can be employed to produce seeds with the chemical composition of the oil, enriched with specific substances that can be used as an alternative therapy in the treatment of various diseases. This chapter will present the nutritional composition and healthy benefits of soybean oil and some potential commercial implications.

Chapter 2 – Chia (*Salvia hispanica* L.) seed oil is a very interesting source with regard to provide a good equilibrium between two essential fatty acids (FAs) (linoleic and α-linolenic acid). Currently, chia seed oil is not widely used commercially even though its characteristics are well-suited for industrial applications, and contribute to healthy human diets. One of the main objectives of chia oil production involves the appropriate selection of the extraction process. The yield and the quality of oil are very important to determine the feasibility of commercial production. Chia seed oil was obtained by different extraction processes, some of them commonly used by the oil industry (solid-liquid extraction and cold pressing) or by alternative technologies with supercritical CO_2 (SC - CO_2). The aim of this work was to analyze the oil yield, the fatty acid composition, the total tocopherol and polyphenolic compounds content and the oxidative stability of chia seed oils obtained by solvent, pressing and CO_2 supercritical extraction (CO_2-SE) by a multivariate statistical method. The highest oil yield was 0.34 g/g seed (d.b.) obtained by solvent extraction (hexane). It was also possible to achieve similar values by adjusting the operating conditions (pressure, temperature and time of extraction) of the SC-CO_2 process. However, the oil yield reached by pressing was about 30% lower than those obtained by solvent (hexane) and SC-CO_2. The fatty acid composition of oils was similar for the different processes, highlighting the α-linolenic (~65%) and linoleic (~20%) acids content and a low level of saturated acids (~9%). Furthermore, the presence of a moderate amount of bioactive compounds such as tocopherols and polyphenols, was recorded. Multivariate analysis showed that the first three principal components described about 92% of the variance. The features that differentiate the oils obtained by conventional processes from those extracted by CO_2-SE were the presence of larger amounts of oleic and stearic acids, tocopherols and oxidative stability in the former, and the increased quantities of palmitic and linoleic (C18:2) acids and total polyphenol compounds in the latter.

Chapter 3 – Rapeseed oil contains high amounts of bioactive compounds, such as polyphenols, phytosterols, tocopherols and other antioxidants, which play an important role in the prevention and treatment of some chronic diseases and improve immune function. In addition to its use as a food, this oilseed is also a viable option for the production of alternative fuels (biodiesel) due to its high oil content and yield per hectare, as well as the good quality of the extracted oil. Sunflower oil is used as a food and as an emollient in ointments and creams. Sunflower oil is essentially free of linolenic acid compared to soybean and rapeseed oils (3-10%). This provides some increased oxidative stability, but does not furnish valuable omega-3 acids that are necessary for health. Tocopherols are the main compounds with antioxidant properties present in sunflower seeds. In the oil extraction process, the seeds undergo a series of unit operations such as drying, storage, crushing, cleaning, flaking, conditioning, mechanical pressing and extrusion followed by solvent extraction. These processing stages may affect the quality and quantity of the oil extracted. First it is necessary to reduce the moisture content of the seeds for safe storage. The literature shows divergent data on the effect of the process temperature on the oil quality (measured in terms of the acidity value, peroxide index and tocopherol content) of different seeds. Conditioning of the seeds prior to extraction is required to make the oil inside the membranes more accessible to the solvent. Pretreatments such as crushing, hydrothermal treatments and the novel microwave technology are applied to seeds in order to modify or break their structure so as to facilitate the release of the oil. These pretreatments could also affect the release of other minor compounds, such as tocopherols. Another method used to make the release of the oil easier is by enzymatic degradation of the cell wall before and/or during extraction, but the release of bioactive compounds is also affected.

Chapter 4 – One of the factors that markedly affects the characteristics and stability of oil-in-water (O/W) emulsions is the presence of polysaccharides in the aqueous phase. O/W emulsions (20:80 wt/wt) were prepared with refined corn oil and dispersions with ≥0.75% mucilage (6.8 and 18.8% protein content) and 0.1% Tween 80, and they presented very good stability during storage for 120 days at 4±1ºC (backscattering value 78%). The emulsions prepared with mucilage with lower protein (6.8%) and lipid content (0.9%) were more stable. The stability of O/W emulsions (40:60 wt/wt) prepared with canola oil and dispersions of mucilage extracted from locust bean and flax seeds was higher when the mucilage concentration was increased from 0.5 to 1.5% in the aqueous phase of the emulsion, whereas the emulsions formulated with mucilage from fenugreek seeds exhibited high stability (100%) during all the storage time (90 days, 25±1ºC) and for all the mucilage concentrations tested (0.5-1.5%). It was also found that flax mucilage reduces the creaming phase in carrot juices, and helps to stabilize meat products by its interactions with the meat proteins. In general, polysaccharide dispersions increased the viscosity of the aqueous phase of the emulsions, limiting the mobility of the oil droplets in the dispersed phase to migrate, and therefore to flocculate or coalesce. Thus the physical stability of O/W emulsions against gravitational phase separation can be improved with the addition of chia mucilage, given its role as a thickening agent.

Chapter 5 – The different vegetable oils available on the market for human consumption mainly differ in fatty acid composition. Chia, flaxseed and sacha inchi oils, are sources of fatty acid α-linolenic (ω-3) followed by mustard and canola oils, while sunflower, safflower, corn, soybean and black cumin oils present high linoleic acid content (ω-6). Polyunsaturated fatty acids (PUFA) (ω-3, ω-6) are essential compounds commonly found in vegetable oils. They are nutritionally important for good health and are especially beneficial for individuals suffering from coronary heart disease, diabetes, and immune response disorders. FAO/WHO have recommended that the essential ω-6:ω-3 FA balance in the diet should be between 5:1 and 10:1. This can be achieved by mixing or blending two or more different oils in specific proportions to get a desired fatty acid composition. Blending vegetable oils can increase the levels of bioactive lipids and natural antioxidants in their blends and improve the nutritional value at affordable prices. Oil blends has been a common practice in the many countries. Recently, the manufacture and marketing of blended oils containing common and unconventional edible oils are allowed. This article deals primarily about blends of different vegetable oils in order to obtain products with improved essential ratio in fatty acids (ω-6:ω-3), functional properties and oxidative stability.

Chapter 6 – Nowadays Jatrophacurcas is one of the important alternative oil plants to produce biodiesel. But because of toxic substance especially phorbol esters are dangerous compounds for human who working with this oil. And so it need to eliminate this substance before utilization.

Phorbol esters are a natural toxic ester found in tropical plant in the family of Euphorbiaceae. It is main toxic compounds in seed oil of Jatrophacurcas. The biological effects of phorbol esters are tumor promotion or cocarcinogen when taken and inflammation when contacted. At least 5 types of phorbol esters are detected in J. curcas oil. The major chemical structure of detected phorbol ester is 12-Deoxy-16-hydroxyphorbol-4'-[12',14'-butadienyl]-6'-[16',18'20-nonatrie-nyl]-bicyclo[3.1.6]hexane-(13-0)-2'-[carboxylate]-(16-0)-3'-[8'-butenoic-10'] ate or DHPB.

Many researchers tried to detoxify phorbol esters in seed oil by the extraction with ethanol or methanol but this experiment is difficult to apply for industrial scale because of the immense solvent consumption. Some researcher studied on tradition oil refining process by using deacidification followed bleaching step. The result of experiment showed only 55% of phorbol esters were removed. So in our experiment, the adsorption technique using bentonite was applied to adsorpphorbol esters compounds. The result showed that the optimum adsorption condition on J. curcas oil was 3.2%(w/v) of bentonite, 15 min of adsorption time, 100 rpm of stirring rate at room temperature. The phorbol esters can be removed up to 98% for one time of adsorption. This technique is recommended for detoxification J. curcas oil in large scale production. In addition, our study also develop a technique to confirm the presence of phobol esters left in oil after adsorption using liquid chromatography-tandem mass spectrometry with multiple reaction monitoring mode that detects the ionization of parent molecule with mass 711 to precursor and product ion with mass 311 and 293 respectively. This technique is useful technique to confirm phorbol esters left in oil.

Chapter 7 – Sinusoidal obstruction syndrome (SOS), previously known as veno-occlusive disease (VOD), occurs in patients undergoing hematopoietic cell transplantation and chemotherapy. SOS is historically called Gulran disease in Afghanistan and senecio disease in South Africa; it dates back to 1920. Pyrrolizidine alkaloids (PAs) in herbal preparations such as tea and Chinese medicine induce SOS. PAs in grasses and animal feed cause acute and chronic poisoning in cattle. The chemotherapeutic drugs oxaliplatin and cyclophosphamide also cause SOS. The search for a novel and effective therapy for chemotherapeutic-drug-induced-SOS continues. Sesame oil is a nutrient-rich antioxidant popular in alternative medicine and traditional health foods in Asian countries. Sesame oil and its lignan sesamol have been proved effective for treating various drug-induced and chemically induced liver injuries. Sesame oil and sesamol maintain glutathione and reduce myeloperoxidase activity, nitrate content, lipid peroxidation (LPO), and the recruitment of inflammatory cells in SOS. In addition, they downregulate matrix metalloproteinase (MMP)-9 expression and upregulate tissue inhibitor of metalloproteinases (TIMP)-1, laminin, and collagen in SOS. The authors hypothesize that sesame oil and sesamol would be useful for treating PA-mimicking chemotherapeutic drug-associated SOS.

Chapter 8 – Nonalcoholic fatty liver disease (NAFLD) is highly prevalent in the general population. Nonalcoholic steatohepatitis (NASH), also called *nutritional steatohepatitis* and *nutritional fibrosing steatohepatitis*, can progress to liver failure and hepatocellular carcinoma. Nutritional fibrosing steatohepatitis has been called "a tale of a two-hit hypothesis": the "First Hit" is characterized by hepatic injury and fat accumulation, and the "Second Hit" is characterized by hepatic oxidative stress, inflammation, and insulin resistance. Managing NAFLD focuses particularly on diet and exercise; managing NASH focuses on lifestyle modifications, control of associated metabolic issues, and pharmacological therapy for liver injury.

Successful care and treatment require an integrative approach. Proposed pharmacological therapies for NASH include vitamin E, ursodeoxycholic acid (a drug used to dissolve gallstones), pioglitazone (one of a class of drugs called thiazolidinediones that are used to treat type 2 diabetes), and metformin (used to treat type 2 diabetes); however, drugs are therapeutically limited and may produce adverse effects. The search for a novel and effective medication to treat nutritional steatohepatitis continues. Sesame oil is nontoxic, antioxidant-rich, and nutritional oil, and it is effective against various diseases models; it attenuates both the first and second hits of nutritional steatohepatitis. Sesame oil attenuates hepatic injury and steatosis, reduces levels of triglycerides, nitric oxide, malondialdehyde (a biomarker of lipid peroxidation), tumor necrosis factor-α, interleukin-6, interleukin-1β, leptin, tissue growth factor-β1, α-smooth muscle actin, fibrosis, and the activity of matrix metalloproteinase-2 and -9, but it increases tissue inhibitor of metalloproteinases-1 and peroxisomal proliferator-activated receptor-γ expression. Thus, the authors hypothesize that sesame oil would be useful for treating NASH.

In: Seed Oil
Editor: Alexis Varnham

ISBN: 978-1-63463-056-6

Chapter 1

SOYBEAN SEED OIL: NUTRITIONAL COMPOSITION, HEALTHY BENEFITS AND COMMERCIAL APPLICATIONS

Luiz Gustavo de Almeida Chuffa**[*,1]**,
Fabrício Rocha Vieira**[2]**,
***Daniela Alessandra Fossato da Silva**[3]*
***and Danilo Miralha Franco**[4]*
[1]Department of Anatomy - IBB/UNESP – Univ. Estadual Paulista, SP.
[2]Department of Plant Production – FCA/UNESP – Univ. Estadual Paulista, SP.
[3]Institute of Biosciences - Univ. Estadual Paulista, SP.
[4]Department of Botany - IBB/UNESP – Univ. Estadual Paulista, SP.

ABSTRACT

The vegetable oils are dietary sources of sterols, vitamin E, and unsaturated fatty acids. The fatty acid composition and the content of an unsaponifiable substance in oilseeds depend on the variety of plant,

* Corresponding author: Luiz Gustavo de Almeida Chuffa. E-mail: guchuffa@yahoo.com.br. Department of Anatomy, Bioscience Institute, Univ. Estadual Paulista. Address: Distrito de Rubião Júnior s/n, Botucatu - São Paulo/SP.

degree of ripening seeds and the climatic conditions. Vegetable soybean *(Glycine max* (L.) Merr.) is an annual dicotyledonous plant belonging to the family of Fabaceae (Leguminosae) in the genus; their genotypes are divided into two categories: large-seeded and small-seeded. The large-seeded types are used for the fresh market in urban areas, whereas small-seeded types are used to make soybean sprouts. There are several groups of nutrients in soybean that are currently under investigation for healthy benefits, such as flavonoids and isoflavonoids, phenolic acids, phytoalexins, phytosterols, proteins and peptides, and saponins. In addition, soybeans are also an important source of minerals copper, manganese, molybdenum, phosphorus, potassium, B vitamin, and omega-3 fatty acids (alpha-linolenic acid). Replacing meat and dairy with soy may decrease total cholesterol intake by about 123 mg/day and saturated fat by about 2.4 g/day. These changes would reduce the risk of developing cardiovascular diseases, and even the cancer. Soybean oil is the edible oil extracted from soybean seeds, largely used as cooking oil in the world according to the US agricultural services. Dry soybean seeds compose 18-20% of extractable oil by weight. The seeds are then subject to pressing to obtain oil and the residue is used as animal feed. The crude soybean oil is yellow in color and contains moisture, lecithin, free-fatty acids, and some volatile compounds. These impurities have to be removed to obtain acceptable standard oil. Soybean oil has a good lipid profile with saturated, monounsaturated and polyunsaturated fats in healthy proportions (SFA:MUFA:PUFA=16:24:58). Linoleic acid (omega-6) is the major polyunsaturated fatty acid found in oil; phytosterols, especially B-sitosterol, inhibit cholesterol absorption and reduce blood LDL-cholesterol levels by 10% to 15%; anti-oxidant vitamin E, a powerful lipid soluble vitamin, is important to maintain the integrity of cell membranes and protect them from harmful reactive oxygen-free radicals; vitamin K, an essential element in promoting bone formation and strengthening, and neuronal protection in the brain.

More recently, genetic engineering is becoming an important tool to produce different and exotic oils derived from a diversity of plants in domesticated and commercial seeds-like soybean, by using genetic transformation. This technique has attracted commercial interest and can be employed to produce seeds with the chemical composition of the oil, enriched with specific substances that can be used as an alternative therapy in the treatment of various diseases. This chapter will present the nutritional composition and healthy benefits of soybean oil and some potential commercial implications.

INTRODUCTION

Soybean Oil: A Brief Overview

Vegetable soybean (*Glycine max* (L.) Merr.) also soya or soja bean, formerly classified as Glycine soja, is an herbaceous plant from the Fabaceae family (legume) naturally originated in southeastern Asia (Japan, korea, and China) that was domesticated 3.000 years ago because of its young pods and edible seeds. Soybean is the world's most important legume crop, and the most widely commercialized oilseed growing in different climates worldwide (Pavlova, 1989). There are two genotyped categories: large- and small-seeded. The large-seeded type are mainly used for the fresh market in urban areas to oriental populations, whereas the small-seeded types are used to prepare soybean sprouts.

Soybean seeds borne on different nodes of the stem and are subjected to positional effects (Bennett et al. 2003). Oil content and fatty acid composition vary between positions along the axis (Guleria et al. 2008). Seeds in the upper one fourth of the plant contain a higher concentration of protein and lower concentration of oil than that from lower one fourth of the plant (Escalante and Wilcox 1993). These differences in the oil and protein availability is due to the variations occurring in specific nutrients and assimilates supply and other related factors, probably influencing the germination of the seed (Sharma et al. 2009).

Soybeans have high amount of protein and oil, and they are used into diverse food products, including soy curd and fermented soy cakes (tofu and tempeh), soy sauce, soy paste (miso), and soy milk. Such hydrolyzed protein is a meat substitute used for many people. Flour made from soybeans is utilized in processed foods, as a stabilizer and to increase protein content. Soy oil is used in cooking, such as margarine, shortening, salad oil as well as in industrial products (paints, printing inks, disinfectants, biofuel, and linoleum). The soy derivatives that remains after oil extraction is used to produce fiber, textiles, adhesives, and livestock feed (Ecocrop, 2012, Wyk, 2005).

The concentration of soybean oil ranges from 83 g/kg to 279 g/kg (Wilson, 2004). As demonstrated in Table 1, there are several different fatty acids present in the soy oil. Soybean oil contains a high amount of unsaturated acids important in the human nutrition: α-linolenic acid (omega-3 acid), linoleic, γ-linolenic and arachidonic acid (omega-6 acid), and oleic acid known as omega-9 (Nikolic et al. 2009, Olguin et al. 2003). There are variations in the components of soybean oils among different samples: the proportion of linolenic acid in the seed oil of ATAEM7 is the highest (53.5%); oleic acid contents of seed oils varied from 21.4% (ATAEM7) to 26.7% (Turksoy).

Table 1. Mean composition of the most lipid content present in the soybean seed oil

Fatty Acid	Soy Oil (%)	% in the glyceride	Commercial soy oil (%)	Melting point (°C)
C4:0	--	--	--	- 8.0
C6:0	--	--	--	- 3.0
C8:0	--	--	--	16.5
C10:0	--	--	--	31.0
C12:0	--	--	0.42	44.0
C14:0	1	< 0.5	0.39	54.0
C16:0	10	7.0 - 14.0	16.43	63.0
C16:1(9)	1	< 0.5	0.14	0.0
C18:0	2	1.4 - 5.5	4.14	70.0
C18:1(9)	28	19.0 - 30.0	18.37	13.0
C18:2 (9,12)	50	44.0 - 62.0	52.80	- 5.0
C18:3(9,12,15)	8	4.0 - 11.0	4.33	- 11.0
C20:0	< 1.0	< 1.0	0.30	75.0
C20:1	0.1 - 0.3	--	--	23.0
C22:0	0.3 - 0.7	--	--	80.0
C22:1	0.3 (max.)	--	--	33.0
C24:0	0.4 (max.)	--	--	84.2

The proportion of linoleic acid of soybean oil ranged from 49% (Turksoy) to 53.5% (ATAEM7), and the palmitic acid of oils varied between 9.2% (Adasoy) and 11.2% (Noya). The major sources of tocopherols were ¥-tocopherol, α-tocopherol, and δ-tocopherol in all varieties of soybean oil, and γ-tocopherol proportion was high in the ATAEM7 (Matthaus and Ozcan, 2014). Stearic acid content in soybean typically represents 2-5% of total fatty acids; however, some germplasm lines have been developed with high stearic acid. These have been developed using mutagenesis, with the exception of FAM94-41 (Pantalone et al. 2002). Specifically, FAM94-41 has a spontaneously mutation in the SACPD-C gene, a seed isoform of a D9-stearoyl-acyl carrier protein-desaturase, which produces stearic phenotype (up to 9%, Zhang et al. 2008, Ruddle et al. 2013). These differences of bioactive compounds of soybean cultivars may be due to soil properties, genetic factors, and growth conditions. Although these acids have reportedly been implicated in reducing cholesterol fractions and associated diseases in humans, they have a negative impact on flavour and oil stability during frying (Kris-Etherton et al. 1993, O'Brien et al. 2005).

The longevity of seed in storage is influenced by the stored quality and conditions. Regardless of initial seed quality, the unfavorable storage conditions (variation in air temperature and humidity) may contribute to accelerating seed deterioration. Besides the storage is associated with a decline in phospholipids and long-chain fatty acids, auto-oxidation of lipids and high content of free fatty acids are the main reasons for deterioration of oil seeds (Balesevic-Tubic et al. 2005). The ageing is related to accumulation of superoxide radical, hydrogen peroxide, and hydroxyl radical that can interact with cellular membranes, enzymes and nucleic acids (Sharma et al. 2006). Otherwise, the protective mechanisms within the seeds include several antioxidant enzymes such as superoxide dismutase, catalase, and glutathione peroxidase (Sung, 1996).

Soybean oil is one of the main oils that contains high amounts of monounsaturated and polyunsaturated fatty acids (MUFA and PUFA). This specific fatty acid composition helps to reduce blood cholesterol fractions, thus lowering the risk of heart disease. However, soybean oil is highly susceptible to oxidative process (Naz et al. 2005). In order to increase the oxidative stability of soybean oil, antioxidants, such as butylated hydroxyanisole, tertbutylhydroquinone (TBHQ), and butylated hydroxytoluene (BHT) have been widely used as food additives (Eshghi et al. 2014).

To investigate the soybean oil oxidation, three different aspects have to be addressed under proper conditions: peroxide value (PV), acid value (AV) and iodine value (IV). Hydroperoxide is the primary product derived from lipid oxidation, whereas their sub products are mostly responsible for rancid undesirable flavour. Eshghi et al. (2014) have reported that when oil samples are maintained at both temperatures (25ºC and 55ºC) in darkness and light, a progressive increase in PVs throughout the storage period is higher than the samples containing antioxidants. They concluded that temperature is a main factor affecting the oil oxidation rate, which starts with the abstraction of hydrogen adjacent of a double bond in a fatty acid to form a free radical.

Soy oil contains about 50% linoleic acid, and recently, Jain and Proctor (2006) proposed a simple way of producing high levels of conjugated linoleic acid (CLA) on a lab scale by converting soy oil linoleic acid to CLA using a UV lamp with 0.15% iodine. CLAs are positional and geometric isomers of octadecadienoic acid, and their double bonds are conjugated and not methylene interrupted as evidenced in linoleic acid (18:2n-6). This form can be found naturally in beef products at levels of 0.3-0.8% (w/w) of the fat content. Curiously, photoirradiation method is able to produce 20% CLA in approximately 144 h. So, Jain et al. (2008) optimized the process on a pilot scale further resulting in greater quantities of CLA in less time (75% of total CLA were *trans, trans-isomers*, and the remaining were *cis, trans- and cis-trans-isomers*). The effect of soy oil components on CLAs yields and oxidative stability during photoisomerization of linoleic acid was determined by Tokle et al. (2009). All of the peroxides, free fatty acids, phospholipids, and lutein had reduced CLA yields, with peroxides having a greatest effect (a PV from 0 to 3-4 mequiv/kg reduces CLA yield by 50%). In contrast, only a 1400 ppm of mixed soy tocopherols produced an increase in CLA yields. Notably, studies on commercial antioxidants and specific tocopherols on CLA yields and oxidative stability during linoleic acid photoisomerization are still needed (Yettella et al. 2011). Interestingly, the stearidonic acid-enriched soybean oil as a dietary component incorporated into baked bars and beverages (7.0 g/day stearidonic acid soy oil/ 1.6 g/day stearidonic acid) increased Omega-3 levels by raising the percentage of eicosapentaenoic acid in red blood cells of healthy men and women. The enrichment of stearidonic acid in soybeans provides an important dietary source of stearidonic acid into a wide variety of food and has greater stability during storage and food production (Whittinghill and Welsby, 2010). Figure 1 illustrates the beneficial effects of the soybean oil-derived compounds in different body systems and diseases.

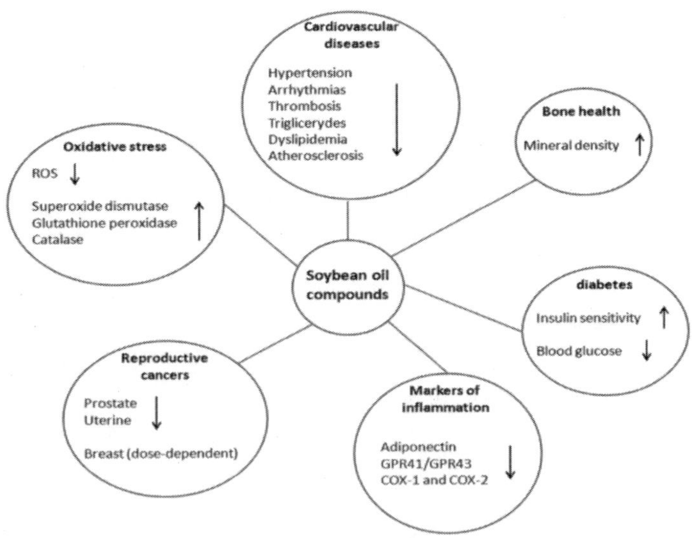

Figure 1. Potential effects of soybean oil on metabolic diseases and related disorders: An overview of several major actions.

Soybean Oil and Healthy Benefits

MUFA and PUFA and Inflammation

High fat intake has been associated with adipose tissue inflammation (Weisberg et al. 2003, Todoric et al. 2006), and furthermore, low-, moderate- and high-fat diets have suggested that the amount of fat present in the diet is less important than the nature of fat consumed (Field et al. 2007, Mozaffarian et al. 2011). Various studies have assessed whether diets containing certain proportions of MUFA and PUFA might affect obesity parameters. While PUFA (n-3) is believed to be beneficial to health, high intakes of PUFA (n-6) without a concurrent increase in PUFA (n-3) have showed detrimental effects on cardiovascular events and even death (Ramsden et al. 2010, 2013). Many changes in fatty synthase, adiponectin metabolism, and short-chain fatty acid receptors GPR41 and GPR43 have been observed in animals fed high-fat diets with different amounts of MUFA and PUFA (Enns et al. 2014). These findings indicate that different types of fatty acid-rich diets (e.g. soybean oil) regulate adipokines and proteins involved in adipose tissue metabolism and inflammation.

Soybean Oil and Diabetes

Insulin resistance is one of the metabolic alterations characterized by an abnormal response of circulating insulin which is associated with glucose intolerance and decreased glucose uptake by peripheral insulin responsive tissues. Therefore, insulin resistance precedes the onset of type 2 diabetes mellitus (T2DM) (Bonadonna and De Fronzo, 1991). Mono- or poly-unsaturated fatty acids have been described to improve insulin sensitivity (Coll et al. 2008) and to exert anti-obesity action (Sekiya et al. 2003). Furthermore, shifting from a diet rich in saturated fatty acids to one rich in mono-unsaturated fatty acids ameliorates insulin sensitivity in healthy people (Bonadonna and De Fronzo, 1991). A recent study showed that animals treated with 100 μL soybean oil for 7 days developed insulin resistance, and the expression of GLUT4, a transmembrane glucose transporter, was 60% lower in adipose tissue while no effects were observed in skeletal muscle (Poletto et al. 2010). It has been shown that low amount of soybean oil, rich in both linoleic and alpha linolenic acids (LA and ALA), ameliorates the diabetic phenotype and restores Δ6 desaturase levels (Leikin Frenkel et al. 2004).

In contrast, an experimental study found that high-fat diet containing soybean oil is able to increase body mass, length, and retroperitoneal fat mass, when administered during the first 60-days-old. In addition, a severe reduction of the pancreatic islets area was observed in these animals (da Costa et al. 2013).

A study with normotensive healthy subjects found that soybean oil, olive oil, and lipid free similarly increased glucose and insulin concentrations during parenteral infusion. However, no changes were observed for lipid profile, inflammatory and oxidative stress biomarkers, or immune functions between soybean oil- and olive oil-based lipid emulsions (Siqueira et al. 2011).

Soybean Oil and Cardiovascular Disease

Soybean oil can be enriched with (n-3) stearidonic acid (SDA) to allow efficient conversion to longer chain eicosapentaenoic acid (EPA). EPA possess distinct biological properties that generally impart properties to cells and tissue, which underlie their ability to promote health and prevent disease (Deckelbaum and Torrejon, 2012). Although active in some areas of human biology, mechanisms of EPA actions are perhaps best defined in cardiovascular disease. The long chain EPA can alter cell membrane structure and fluidity as well as decreasing the amount of membrane occupied by lipid rafts (Yaqoob and Shaikh, 2010). The (n-3) fatty acid (FA) and their derivatives are important molecules in chemotaxis and immune and inflammatory response. Importantly, (n-3) FA decrease blood pressure and alter vascular resistance (Sudheendran et al. 2010). Some cardiovascular benefits have been reported for (n-3) FA in terms of reducing arrhythmias, providing TG-lowering effects, and providing antithrombotic and antiinflammatory as well as antihypertensive effects (Sudheendran et al. 2010). In a variety of experimental studies, animal models fed a high-fat diets rich in (n-3) FA decreased dyslipidemia, cholesterol delivery to the arterial wall, arterial proinflammatory processes, and increased arterial anti-inflammatory biomarkers. Also, LDL uptake can be affected by different types of dietary FA (Rumsey et al. 1992), in which the uptake of cholesterol esters from LDL can lead to cholesterol deposition within cells and tissues, thereby contributing to atherosclerosis. Satisfactorily, high-fat diets rich in (n-3) FA negatively regulate selective uptake and decrease whole-particle LDL uptake, with a severe LDL reduction in the aortic media layer (Chang et al. 2009).

Growing evidences have pointed out that oxidative stress leads to vascular damage and plays a critical role in the cardiovascular diseases such as hypertension (Fearon and Faux, 2009). High production of reactive oxygen species (ROS) is increased during hypertension in both experimental and clinical models of hypertension (de Champlain et al. 2004). After comparing the effects of canola oil and soybean oil ingestion upon antioxidant activities,

Papazzo et al. (2011) reported that soybean oil promoted elevation in superoxide dismutase, glutathione peroxidase and catalase activities, total cholesterol and low-density lipoprotein cholesterol. In contrast, malondialdehyde and 8-isoprostane concentrations were significantly lower after the ingestion of canola oil compared to the soybean oil.

Parenteral infusion of Soybean oil significantly reduced brachial artery flow-mediated dilatation from baseline - 23% at 4 h and - 25% at 24 h; in contrast, administration of olive oil, lipid free, and saline did not change either systolic blood pressure or flow (Siqueira et al. 2011).

Soybean and Phytoestrogens

Isoflavones are active compounds in soy-derived foods. Isoflavones acts in the body similarly to estrogen as they are called phytoestrogens. The most common phytoestrogenic isoflavones are genistein, daidzein, and glycitein (Tsangalis et al. 2005, Valachovicova et al. 2004). Because many women have demonstrated an increased risk of breast cancer, stroke, and heart attacks in response to estrogen and progesterone treatments, the isoflavones intake has become a popular alternative to estrogen therapy (Messina, 2002). By binding to the estrogen receptors (ER) of the cells, isoflavones produces weak estrogenic effects, mainly when an inadequate amount of estrogen is present in the body. As isoflavones play these double-edge functions in the body, they may help in preventing osteoporosis and menopausal symptoms as well as reducing the risk of breast and uterine cancers by blocking the ER activation (Song et al. 2007).

An study conducted in 177 postmenopausal women who consumed isoflavones extract equivalent to 50 mg/day for 12 weeks showed a reduction in self-reported hot flash severity but not in hot flash incidences (Upmalis et a. 2007). In contrast, Nikander et al. (2003) who administered a daily 114 mg of isoflavones for 12 weeks, and Penotti et al. (2003) who administered daily 72 mg of isoflavones for 6 months, observed no differences between the treatments and placebo groups. These inconsistencies are probably due to individual variability, measurements index, duration of treatment, dosages, and isoflavones properties. In general, the beneficial effects on menopausal symptoms were more prominent in studies that used isoflavones supplements rather than soy proteins.

Bone health is a major concern in elderly women. Curiously, studies with Asiatic women (China) who consumed the most soy-derived foods were one-third less likely to acquire a bone fracture that Chinese women who consumed the lowest amount of soy (Zhang et al. 2005).

This fact has led to the hypothesis that soybean isoflavones represent an alternative option for the prevention of osteoporosis. Three controlled studies (Chen et al. 2003, Morabito et al. 2002) with isoflavone extracts or genistein demonstrated that soy isoflavones have a mild, but significant effect on the improvement of bone mineral density at doses ranging between 35 to 54 mg of aglycone, while other studies showed no effect with doses ranging from 4 to 103 mg of aglycone equivalents (Gallagher et al. 2004, Kreijkamp-kaspers et al. 2004). Additional studies are needed to better understand the real effect of isoflavones on bone structure.

Soy consumption has been associated with lower risks of developing cancer or tumor (Kim et al. 2004, Sarkar and Li, 2003, Stoll, 1997). Prostate cancer is known to have increased levels of dihydrotestosterone (DHT), and isoflavones inhibited 5alpha-reductase, which is involved in the conversion of testosterone to DHT (Yi et al. 2002).

Although a small number of epidemiologic studies support a negative correlation between soy isoflavones and breast cancer risk, many of the case-control studies pointed to important limitations (Trock et al. 2006, Yan and Spitznagel, 2005).

Supplementation with isoflavones at 43.5 mg/day for 12 months or 85.5 mg/day for 6 months (Atkinson et al. 2004) produced no changes in mammography density, serum estradiol, follicle stimulating hormone, and luteinizing hormone levels in postmenopausal women. While isoflavones have anti-estrogenic effect by blocking endogenous estrogen, experimental data or cultured human breast cell lines showed evidence that soy isoflavones might stimulate breast cancer cells (Martin et al. 1978, Hsich et al. 1998).

In general, isoflavone can stimulate the breast cancer cells in women who had already developed breast cancer (a subtype of estrogen-dependent tumor). High levels of estrogens stimulate uterine cells and increase the risk of uterine cancer in susceptible individuals. There are several evidences showing that isoflavones have no effect on the growth of uterine cells (Crisafulli et al. 2004, Wood et al. 2004, Nikander et al. 2005). However, high doses of isoflavones, specially genistein, stimulated uterine growth and expression of estrogen-regulated genes in uterus (Diel et al. 2001).

Lastly, the risk for postmenopausal women to develop an endometrial cancer under a chronic consumption of soybean is reported to be low.

Commercial Applications

Application and Extraction methods

Currently, soybean is the most cultivated oilseed crop in the world with most producers' concentrated in Americas and Asia regions (Fargione et al. 2008). In 2013, the largest soybean grain and oil producers are the United States, Brazil, Argentina, China, India and Paraguay, which represent over 50% all of soybeans produced in the world (Table 2; Faostat, 2014). In the last decades, global increases of protein and soybean oil per ha cultivated were achieved through the selection of genotypes with high productivity and sophisticated farming techniques genotypes (Clemente and Cahoon, 2009). The current outlook of soybeans is totally favorable an expanded use of soybean oil as a renewable chemical feedstock. However, the increased quantity of protein and oil didn't bring benefits to the physical and chemical properties of the oil that still have limitations and implications for many applications of soybean oil in the food industry (Mahmoud et al. 2006). Some implications are related to the content of palmitic acid, stearic acid, oleic acid, linoleic acid, which show low oxidative instability and different functionalities (Clemente and Cahoon, 2009).

Table 2. World soybean production in grains and oils (2013)

Country	Production (million tons)	
	Production (grains)	Oil production
United Sates	82.05	9.20
Brazil	65.85	6.92
Argentina	40.10	6.35
China	12.80	10.66
India	11.50	1.60
Paraguay	8.35	0.21
All others	34.02	17.25
Total	**254.68**	**52.18**

Source: Food and Agriculture Organization of the United Nations – FAO (June, 2014). Author's calculation.

Soybean oil and protein are the major economic products obtained from soybean seed (Fargione et al. 2008). Soybean oil represents 56% of total oilseed production in the world and is the second most consumed with 27%

(SoyStats, 2014). In the US, most of soybean oil productions are used in foods and food processing annually, and 4% of total soybean oil is used in non-edible industry for production of fatty acids, soaps, animal feed, manufacture of inks, paints, varnishes, resins, plastics, and biodiesel (Cahoon, 2003). Nowadays dietary trend is to gradually replace animal fat to vegetable fat, including countries where people traditionally consume fat predominantly from animals (Tuberoso et al. 2007). The residue of soybean oil extraction referred to, as soybean meal is the most important source of protein used to feed farm animals (Oil World, 2010). This bioproductof soybean represents two thirds of the global total protein food for commercial animals (Oil World, 2010, Faostat, 2014). This is due to two points, high protein content of 44 to 50%, its consistent availability and constantly competitive price (Steinfeld et al. 2006). The biodiesel is another product derived from soybean crude oil that is being appointed as energy source usable in the world (Fargione et al. 2008). The soybean oil has been applied to produce effective insect repellents against arthropod-transmitted diseases (Fradin and Day, 2002). In printing industry, the soybean oil are helping the industry in reducing the environmental burden of the printing industry beyond widely available at low cost. In health therapies, soybean oil has been used as nutritional supplement in intravenous feedings and lipid emulsions. The administration of parenteral nutrition containing soybean oil-based and olive oil-based lipid emulsion resulted in similar rates of infectious and noninfectious complications, and no differences in glycemic control, inflammatory and oxidative stress markers, and immune function has been described in critically ill adults (Umpierrez et al. 2013). Preparation of avocado–soybean unsaponifiables (ASU) in osteoarthritis (OA) patients was tested and the results suggest that ASU are no worse and no better for treatment of OA than other medicaments (Christensen et al. 2008).

Seed oils are described as composed majority by triacylglycerides, but contains a variety of other components, such as diacylglycerides, monoacylglycerides, tocopherol, pigments, phospholipids, free fatty acids, phytosterols, hydrocarbons and water (Chen, McClements and Decker, 2014). Triacylglycerides are esters of glycerol and are produced by the esterification reaction; on the other hand, the hydrolysis of triglycerides produces glycerol and fatty acids (Kimura et al. 1983).

Edible oilseeds are generally obtained through mechanical pressing and solvent extraction (Sawada et al. 2014). In solvent extraction, hexane is the most used for oilseed extraction, but the use of this solvent safety presents implications surrounding the use of hexane (Rosenthal et al. 1996). The use of alternative solvents such as isopropanol and ethanol, and supercritical carbon dioxide has increased recently due to concerns of environmental health and safety (Dunnuck, 1991). However, different solvents present different affinity to solute and substances, and compared to hexane, alcoholic and supercritical carbon dioxide show a less affinity with solute, but pressure or high-intensity ultrasound may reduce the time required to extract edible oils from plant tissue and improve the production of commercial oil (Li et al. 2004). Another extraction methods described in the literature are related to enzyme-assisted aqueous extraction, using two enzymes, proteases and cellulase, that expose seed content, breaking the bonds of the cell wall to facilitate aqueous extraction. Although these methods have effectively showed a higher yield in the oil extraction, they have a higher cost (Rosenthal et al. 1996, Rosenthal et al. 2001).

The different substances use to extraction may select different classes of compounds in the oil. In this way, different cultivation, extraction and post-extraction methods can be used to different destinations in the food, chemical or biofuels industry in accordance to the needs and specifications for each application.

Genetics Engineering and Transgenic Seeds

Plants lipid composition present a natural variety in different species depending on the environmental conditions, and play an important role in adaptation and tolerance to different types of biotic and abiotic stress (Ohlrogge and Browse, 1995, Marchive et al. 2014). This natural variety of the lipid composition indicate a diversification in the lipid biosynthesis that is correlated to specific genes and enzymes (Broun et al. 1999, Marchive et al. 2014).

Molecular gene transfer is becoming an important tool to produce different and exotic oils derived from a diversity of plants in domesticated and commercial seeds-like soybean. This technique has attracted commercial interest and can be employed to produce seeds with the chemical composition of the oil, enriched with specific substances (Knutzon et al. 1992, Maheshwari and Kovalchuk, 2014).

Generally, vegetable oils are composed majority of unsaturated 18-carbon fatty acids: the monounsaturated oleic, polyunsaturated linoleic and linolenic acids; however, most oils also contain small but significant amounts of the saturated palmitic and stearic acids (Downey, 1983, Wendlinger et al. 2014). Genetic transformation of plant lipids depends on the availability of genes for the enzymes that catalyze lipids synthesis in plants, and genes of fatty acyl desaturases that introduce the double bonds required for synthesis of α and γlinolenate were also identified (Girke et al. 1998). Knutzon et al. (1992) have shown that modification on *Brassica* stearoyl-acyl carrier protein (stearoyl-ACP) desaturase gene resulted in transgenic plants with a great increase in stearate levels in the seed. This is commonly observed because the enzyme stearoyl-ACP catalyzes the initial desaturation reaction in fatty acid biosynthesis, and plays important role in the ratio of total saturated to unsaturated fatty acids in plants (Thompson et al. 1991).

Abbadi et al. (2004) studying the polyunsatured fatty acids in soybean used animal enzymes D5- D6-desaturase to increase the content of (VLCPUFA) stearidonic acid (EPA) and eicosapentaenoic rather than acyl-CoA as substrate. Regarding these enzymes, the conversion of linoleic and a-linoleic acid was accomplished exclusively by the acyl-CoA track, thus avoiding the switching between lipids and acyl-CoA. Another investigation conducted by Burr et al. (2002) suggested that down-regulating expression of FAD2 genes together with genes that control the production of palmitic acid (i.e., FATB genes) was able produce soybean seeds with oleic acid content greater than 85% of the total oil.

The main saturated fatty acids in vegetable oils are palmitate and stearate, and these fatty acids are not wanted to dietary effects or food functionality. Furthermore, there is a growing interest in controlling the relative amounts of these fatty acids in vegetable oils for health issues (Broun et al. 1999, Vaz et al. 2014).

To reduce enzyme activity in plants, a method of gene silencing has been used through antisense or co-suppression technologies, and involves the introduction of a modified gene that produces complementary RNA transcripts to gene to be silenced (Bourque, 1995, Liu and Zhu, 2014). Regulating the relative amounts of stearate and palmitate that may be controlled by the ratio of palmitoyl thioesterase and KASII activities, Kinney (1996) obtained a significant increase in palmitate at the cost of the fatty acids of 18 carbons by decreasing the amount of soybean KASII after co-suppression.

Therefore, the discovery of genomes and metabolic pathways for the production of lipids with commercial interest will enable diversification of the use of vegetable oils for different purposes, or to oils having higher production of interest compounds for processing functional foods, or even to the production of biofuels and other non-edible products derived from oilseed.

CONCLUSION

Many information from animal and *in vitro* studies have provided plausible mechanisms to explain how soy-derived foods may influence healthy and nutritional status. While some data indicate soybean to have beneficial effects, such as reducing blood cholesterol, diabetes, oxidative process, and inflammation, others found no effect on body systems.

These contradictory studies is mainly due to the lack of scientific data from well-designed experimental, clinical, and epidemiologic studies. Furthermore, their biological effects are dependent on many factors including dose, time exposure, protein binding affinity, metabolism, estrogen fluctuations, and different action of these compounds in specific population groups. To date, studies on the nutritional quality of immature seeds of soybeans at reproductive stages to determine a more appropriate stage that can be used for human consumption remain a matter of debate. It is essential to differentiate the health effect attributed to pharmacological doses from those physiological or nutritional doses. Pharmacological dose is clinically used to treat diseases and may require a doctor's prescription; however, either physiological or nutritional doses are used to maintain or improve health quality, such as recommended in dietary supplements. Changes in the composition of soybean oil have been satisfactorily modified through genetic engineering and biotechnology. Some advances have demonstrated to be able to alter both qualitative and quantitative composition of soybean oil. However, these techniques are too expensive and require long time to give results in the field. Beyond the genetic tools, other way to ensure the oil quality is the improvement of extraction method with useful possibilities for human health.

REFERENCES

Abbadi, A., Domergue, F., Bauer, J., Napier J.A., Welti, R., Zähringer, U., Cirpus, P., Heinz, E. (2004). Biosynthesis of very-long-chain polyunsaturated fatty acids in transgenic oilseeds: constraints on their accumulation. *Plant Cell*, 16, 2734-2748.

Atkinson, C., Warren, R.M., Sala, E., Dowsett, M., Dunning, A.M., Healey, C.S., Runswick, S., Day, N.E. & Bingham, S.A. (2004). Red-clover-derived isoflavones and mammographic breast density: a double-blind, randomized, placebo-controlled trial [ISRCTN42940165]. *Breast Cancer* Res, 6, R170–R179.

Balesevic-Tubic, S., Malencic, D., Talic, M., Miladinovic, J. (2005). Influence of ageing process on biochemical changes in sunflower seeds. *Helia,* 28, 107–114.

Bennett, J.O., Krishnan, A.H., Weibold, W.J., Krishnan, H.B. (2003). Positional effect on protein and oil content and composition of soybeans. *J Agric Food Chem*, 51, 6882–6886.

Bonadonna, R.C. & De Fronzo, R.A. (1991). Glucose metabolism in obesity and type 2 diabetes. *Diabete Metab*, 17, 112–135.

Bourque, J.E. (1995). Antisense strategies for genetic manipulations in plants. *Plant. Sci*., 105, 125–49.

Broun, P., Gettnet, S., and Somerville, C. (1999). Genetic engineering of plant lipids. *Annu. Rev. Nutr*., 19, 197–216.

Buhr, T., Sato, S., Ebrahim, F., Xing, A.Q., Zhou, Y., Mathiesen, M., Schweiger, B., Kinney, A., Staswick, P. & Clemente, T. (2002). Nuclear localization of RNA transcripts down-regu- late seed fatty acid genes in transgenic soybean. *Plant J,* 30, 155-163.

Cahoon, E.B. (2003). Genetic Enhancement of Soybean Oil for Industrial Uses: Prospects and Challenges. *AgBioForu*m, 6, 11-13.

Clemente, T.E. & Cahoon, E.B. (2009). Soybean oil: genetic approaches for modification of functionality and total content. *Plant Physiol*, 151, 1030-1040.

Chen, B., McClements, D.J., & Decker, E.A., (2014). Impact of diacylglycerol and monoacylglycerol on the physical and chemical properties of stripped soybean oil. *Food Chem,* 142, 365-372.

Chang, C.L., Seo, T., Matsuzaki, M., Worgall, T.S. & Deckelbaum, R.J. (2009). n-3 Fatty acids reduce arterial LDL-cholesterol delivery and arterial lipoprotein lipase levels and lipase distribution. *Arterioscler Thromb Vasc Biol*, 29, 555–561.

Chen, Y., Ho, S., Lam, S., Ho, S. & Woo, J. (2003). Soy isoflavones have a favorable effect on bone loss in Chinese postmenopausal women with lower bone mass: a double-blind, randomized, controlled trial. *J Clin Endocrinol Metab,* 88, 4740–4747.

Coll, T., Eyre, E., Rodrıguez-Calvo, R., Palomer, X., Sánchez, R.M., Merlos, M., Laguna, J.C. & Vázquez-Carrera, M. (2008). Oleate reverses palmitate-induced insulin resistance and inflammation in skeletal muscle cells. *J Biol Chem*, 283, 11107–11116.

da Costa, C.A., Carlos, A.S., de Sousa Dos Santos, A., de Moura, E.G. & Nascimento-Saba, C.C. (2013). High-fat diets containing soybean or canola oil affect differently pancreas function of young male rats. *Horm Metab Res,* 45, 652-654.

Christensen, R., Bartels, E.M., Astrup, A. & Bliddal, H. (2008). Symptomatic efficacy of avocado-soybean unsaponifiables (ASU) in osteoarthritis (OA) patients: a meta-analysis of randomized controlled trials. *Osteoarth Cartil,* 16, 399-408.

Diel, P., Smolnikar, K., Schulz, T., Laudenbach-Leschowski, U., Michna, H. & Vollmer, G. (2001). Phytoestrogens and carcinogenesis - differential effects of genistein in experimental models of normal and malignant rat endometrium. *Hum Reprod*, 16, 997–1006.

Downey, R.K. (1983). *High and Low Erucic Acid Rapeseed Oils*, eds. Kramer, J., Sauer, F. & Pigden, W. (Academic, New York), pp. 1-19.

Dunnuck, J. (1991). NTP technical report on the toxicity studies of of n-hexane in b6c3f1 mice. *Toxicity Report Series*. 2, 1–32.

Ecocrop. 2012. iGlycine max. United Nations Food and Agriculture Organization, Ecocrop online database. Retrieved 25 June 2012 from http://ecocrop.fao.org/ecocrop/srv/en/cropView?id=1150.

Enns, J.E., Hanke, D., Park, A., Zahradka, P. & Taylor, C.G. (2014). Diets high in monounsaturated and polyunsaturated fatty acids decrease fatty acid synthase protein levels in adipose tissue but do not alter other markers of adipose function and inflammation in diet-induced obese rats. *Prostaglandins Leukot Essent Fatty Acids*, 90, 77-84.

Escalante, E.E., Wilcox, J.R. (1993). Variation in seed protein among nodes of normal-and high-protein soybean genotypes. *Crop Sci*, 33, 1164–1166.

Eshghi, N., Asnaashari, M., Haddad Khodaparast, M.H. & Hosseini, F. (2014). Evaluating the potential of natural curcumin for oxidative stability of soybean oil. *Nat Proc Res*, in press.

Fargione, J., Hill, J., Tilman, D., Polasky, S., Hawthorne, P. (2008). Land clearing and the biofuel carbon debt. *Science,* 319, 1235-1238.

Faostat 2014. http://faostat.fao.org (2014) (Accessed in: July 20 2014).

Fearon, I.M. & Faux, S.P. (2009). Oxidative stress and cardiovascular disease: Novel tools give (free) radical insight. *J Mol Cell Cardiol*, 47, 372-381.

Field, A.E., Willett, W.C., Lissner, L. & Colditz, G.A. (2007). Dietary fat and weight gain among women in the Nurses' Health Study. *Obesity*, 15, 967–976.

Fradin, M.S., Day, J.F. (2002). Comparative efficacy of insect repellents against mosquito bites. *N Engl J Med*, 347, 13-18.

Gallagher, J., Satpathy, R., Rafferty, K. & Haynatzka, V. (2004). The effect of soy protein isolate on bone metabolism. *Menopause,* 11, 290–298.

Girke, T., Schmidt, H., Zaehringer, U., Reski, R. & Heinz, E. (1998). Identification of a novel 1-6-acyl-group desaturase by targeted gene disruption in Physcomitrella patens. *Plant J*, 5, 39–48.

Guleria, S., Sharma, S., Munshi, S.K. (2007). Compositional changes in soybean seeds influenced by their positions on stem axis. *J Food Sci Technol,* 44, 607–610.

Hsieh, C., Santell, R., Haslam, S. & Helferich, W. (1998). Estrogenic effects of genistein on the growth of estrogen receptor-positive human breast cancer (MCF-7) cells in vitro and in vivo. *Cancer Res*, 58, 3833–3838.

Jain, V.P. & Proctor, A. (2006). Photocatalytic production and processing of conjugated linoleic rich soy oil. *J Agric Food Chem*, 54, 5590–5596.

Jain, V.P., Proctor, A. & Lall, R.K. (2008). Pilot-scale production of conjugated linoleic acid rich soy oil by photoirradiation. *J. Food Sci.* 73, E183–E193.

Kim, H., Hall, P., Smith, M., Kirk, M., Prasain, J.K., Barnes, S. & Grubbs, C. (2004). Chemoprevention by grape seed extract and genistein in carcinogen-induced mammary cancer in rats is diet dependent. *J Nutr,* 134, 3445S-3452S.

Kimura, Y., Tanaka, A., Sonomoto, K., Nihira, T., & Fukui, S. (1983). Application of immobilized lipase to hydrolysis of triacylglyceride. *Eur J Appl Microbiol Biotechnol*, 17, 107-112.

Kinney, A.J. (1996). β-Ketoacyl-ACP synthetase II genes from plants. US Patent 5,500,361.

Knutzon, D.S, Thompson, G.A., Radke, S.E., Johnson, W.B., Knauf, V.C. & Kridl, J.C. (1992). Modification of Brassica seed oil by antisense expression of a stearoyl-acyl carrier protein desaturase gene. *Proc. Natl. Acad. Sci.* 89, 2624-2628.

Kreijkamp-Kaspers, S., Kok, L., Grobbee, D.E., de Haan, E.H., Aleman, A., Lampe, J.W. & van der Schouw, Y.T. (2004). Effect of soy protein containing isoflavones on cognitive function, bone mineral density, and plasma lipids in postmenopausal women: a randomized controlled trial. *JAMA*, 292, 65-74.

Kris-Etherton, P.M., Derr, J., Mitchell, D.C. (1993). The role of fatty acid saturation on plasma lipids, lipoproteins, and apoproteins: I. Effects of whole food diets high in cocoa butter, olive oil, soybean oil, dairy butter, and milk chocolate on the plasma lipids of young men. *Metabolism,* 42, 121–129.

Lemke, S.L., Maki, K.C., Hughes, G., Taylor, M.L., Krul, E.S., Goldstein, D.A., Su, H., Rains, T.M. & Mukherjea, R. (2013). Consumption of stearidonic acid-rich oil in foods increases red blood cell eicosapentaenoic acid. *J Acad Nutr Diet,* 113, 1044-1056.

Martin, P., Horwitz, K., Ryan, D. & McGuire, W. (1978). Phytoestrogen interaction with estrogen receptors in human breast cancer cells. Endocrinology, 103, 1860–1867.

Li, H., Pordesimo, L. & Weiss, J. (2004). High intensity ultrasound-assisted extraction of oil from soybeans, *Food Research International*, 37, 731-738.

Liu, R. & Zhu, J. (2014). Non-coding RNAs as potent tools for crop improvement. *Natl. Sci. Rev*., 1, 186-189.

Maheshwari, P. & Kovalchuk, I. (2014). Genetic engineering of oilseed crops. *Biocatalysis and Agricultural Biotechnology*, 3, 31-37.

Marchive, C., Nikovics, K., To, A., Lepiniec, L. & Baud, S. (2014). Transcriptional regulation of fatty acid production in higher plants: molecular bases and biotechnological outcomes. *Eur. J. Lipid Sci. Technol.,* 1, 1438-9312.

Matthaus, B.,Ozcan, M.M. (2014). Fatty acid and tocopherol contents of several soybean oils. *Nat Prod Res,* 28, 589-592.

Messina, M. (2002). Soy foods and soybean isoflavones and menopausal health. *Nutr Clin Care*, 5, 272–282.

Morabito, N., Crisafulli, A., Vergara, C., Gaudio, A., Lasco, A., Frisina, N., D'Anna, R., Corrado, F., Pizzoleo, M.A., Cincotta, M., Altavilla, D., Ientile, R. & Squadrito, F. (2002). Effects of genistein and hormone-replacement therapy on bone loss in early postmenopausal women: a randomized double blind placebo-controlled study. *J Bone Miner Res*, 17, 1904-1912.

Mozaffarian, D., Hao, T., Rimm, E.B., Willett, W.C. & Hu, F.B. (2011). Changes in diet and lifestyle and long-term weight gain in women and men. *N Engl J Med,* 364, 2392-2404.

Naz, S., Siddiqi, R., Sheikh, H. & Sayeed, S.A. (2005). Deterioration of olive, corn and soybean oils due to air, light, heat and deep frying. *Food Res Int*, 38, 127–134.

Nikander, E., Kilkkinen, A., Metsä-Heikkilä, M., Adlercreutz, H., Pietinen, P., Tiitinen, A. & Ylikorkala, O. (2003). A randomized placebo-controlled crossover trial with phytoestrogens in treatment of menopause in breast cancer patients. *Obstet Gynecol,* 101, 1213-1220.

Nikolic, N.C., Cakic, S.M., Novakovic, S.M., Cvetkovic, M.D., Stankovic, M.Z. (2009). Effect of extraction techniques on yield and composition of soybean oil. *Macedonian Journal of Chemical and Chemical Engineering*, 28, 173–179.

O'Brien, R.D., Lynn, A.J., Clay, K.C., Phillip, J.W., Peter, J.W. (2005). Cotton seed oil. In: F. Shahidi (Ed.), *Bailey's Industrial Oil & Fat Products* (pp. 173–279). Vol. 2, Hoboken, New Jersey: John Wiley & Sons.

Oil World, (2010). Major meals, World summary balances. *Oil World Weekly*. 55. 45.

Ohlrogge, J. & Browse, J. (1995). Lipid biosynthesis. Plant Cell, 7, 957–70.

Olguin, M.C., Hisano, N., D'Ottavio, E.A., Zingale, M.I., Revelant, G.C., Calderari, S.A. (2003). Nutritional and antinutritional aspect of an Argentinean soy flour assessed on weanling rats. *Journal of Food Composition Analysis*, 16, 441–448.

Pantalone, V.R., Wilson, R.F., Novitzky, W.P., Burton, J.W. (2002). Genetic regulation of elevated stearic acid concentration in soybean oil. *J Am Oil Chem Soc*, 79, 549–553.

Papazzo, A., Conlan, X.A., Lexis, L. & Lewandowski, P.A. (2011). Differential effects of dietary canola and soybean oil intake on oxidative stress in stroke-prone spontaneously hypertensive rats. *Lipids Health Dis,* 10, 98.

Pavlova, N.S. (1989). Fabaceae. In Plantae. Orientis Extremi.Vol.4.Leningrad (Rus) van Wyk, B.E. 2005. "Glycine max." *Food Plants of the World: An Illustrated Guide*. Portland, OR: Timber Press. p. 201.

Penotti, M., Fabio, E., Modena, A.B., Rinaldi, M., Omodei, U. & Viganó, P. (2003). Effect of soy-derived isoflavones on hot flushes, endometrial thickness, and the pulsatility index of the uterine and cerebral arteries. *Fertil Steril,* 79, 1112-1117.

Poletto, A.C., Anhe, G.F., Eichler, P., Takahashi, H.K., Furuya, DT, Okamoto, M.M., Curi, R. & Machado, UF. (2010). Soybean and sunflower oil-induced insulin resistance correlates with impaired GLUT4 protein expression and translocation specifically in white adipose tissue. *Cell Biochem Funct,* 28, 114-121.

Ramsden, C.E., Hibbeln, J.R., Majchrzak, S.F. & Davis J.M. (2010). n–6 fatty acid-specific and mixed polyunsaturate dietary interventions have different effects on CHD risk: a meta-analysis of randomised controlled trials. *Br J Nutr,* 104, 1586-1600.

Ramsden, C.E., Zamora, D., Leelarthaepin, B., Majchrzak-Hong, S.F., Faurot, K.R., Suchindran, C.M., Ringel, A., Davis, J.M. & Hibbeln, J.R. (2013). Use of dietary linoleic acid for secondary prevention of coronary heart disease and death: evaluation of recovered data from the Sydney Diet Heart Study and updated meta-analysis. *BMJ,* 346, e8707.

Rosenthal, A., Pyle, D.L. & Niranjan, K. (1996). Aqueous and enzymatic processes edible oil extraction. *Enzyme and Microbial Technology*, 19, 402-420.

Rosenthal, A., Pyle, D.L., Niranjan, K., Gilmour, S. & Trinca, L. (2001). Combined effect of operational variables and enzyme activity on aqueous enzymatic extraction of oil and protein from soybean. *Enzyme and Microbial Technology,* 28, 499-509.

Ruddle, P., Whetten, R., Cardinal, A., Upchurch, R.G., Miranda, L. (2013). Effect of a novel mutation in a D9-stearoyl-ACP-desaturase on soybean seed oil composition. *Theor Appl Genet*, 28, 107-114.

Sarkar, F. & Li, Y. (2003). Soy isoflavones and cancer prevention. *Cancer Invest,* 21, 744–757.

Sawada, M.M., Venâncio, L.L., Toda, T.A. & Rodrigues, C.E.C. (2014). Effects of different alcoholic extraction conditions on soybean oil yield, fatty acid composition and protein solubility of defatted meal. *Food Research International*, 62, 662-670.

Sekiya, M., Yahagi, N., Matsuzaka, T., Najima, Y., Nakakuki, M., Nagai, R., Ishibashi, S., Osuga, J., Yamada, N. & Shimano, H. (2003). Polyunsaturated fatty acids ameliorate hepatic steatosis in obese mice by SREBP-1 suppression. *Hepatology*, 38, 1529–1539.

Sharma, S., Gambhir, S., Munshi, S.K. (2006). Effect of temperature on vigour and biochemical composition of soybean seed during storage. *J Res Punjab Agric Univ,* 41, 34–38.

Sharma, S., Kaur, A., Sital, J.S. (2009). Effect of storage on germinability and composition of seeds from different positions on soybean stem axis. *Ind J Agric Biochem*, 22, 94–97.

Siqueira, J., Smiley, D., Newton, C., Le, N.A., Gosmanov, A.R, Spiegelman, R., Peng, L., Osteen, S.J., Jones, D.P., Quyyumi, A.A., Ziegler, T.R. & Umpierrez, G.E. (2011). Substitution of standard soybean oil with olive oil-based lipid emulsion in parenteral nutrition: comparison of vascular, metabolic, and inflammatory effects. *J Clin Endocrinol Metab*, 96, 3207-3216.

Song, W.O., Chun, O.K., Hwang, I., Shin, H.S., Kim, B.G., Kim, K.S., Lee, S.Y., Shin, D. & Lee, S.G. (2007). Soy isoflavones as safe functional ingredients. *J Med Food*, 10, 571-580.

SoyStats 2014. http://www.soystats.com (2014) (Accessed in: August 13 2014).

Steinfeld, H., Gerber, P., Wassenaar, T., Castel, V., Rosales, M., de Haan, C. (2006). Livestock's long shadow: environmental issues and option. 377p.

Stoll, B. (1997). Eating to beat breast cancer: potential role for soy supplements. *Ann Oncol,* 8, 223–225.

Sung, J.M. (1996). Lipid peroxidation and peroxide scavenging in soybean seeds during ageing. *Physiol Plant*, 97, 85–89.

Thompson, G., Scherer, D.E., Foxall-Van. S., Kenny, J.W., Young, H.L., Shintani, D.K., Kridl, J.C. & Knauf, V.C. (1991). Primary structures of the precursor and mature forms of stearoyl-acyl-carrier-protein desaturase from safflower embryos and requirement of ferredoxin for enzyme activity. *Proc. Natl, Acad. Sci.*, 88, 2578-2582.

Todoric, J., Löffler, M., Huber, J., Bilban, M., Reimers, M., Kadl, A., Zeyda, M., Waldhäusl, W. & Stulnig, T.M. (2006). Adipose tissue inflammation induced by high-fat diet in obese diabetic mice is prevented by n−3 polyunsaturated fatty acids. *Diabetologia*, 49, 2109-2119.

Tokle, T., Jain, V.P. & Proctor, A. (2009). Effect of minorconstituents on soy oil conjugated linoleic acid production. *J Agric Food Chem*, 57, 8989–8997.

Trock, B., Hilakivi-Clarke, L. & Clarke, R. (2006). Meta-analysis of soy intake and breast cancer risk. *J Natl Cancer Inst,* 98, 459–471.

Tsangalis, D., Wilcox, G., Shah, N. & Stojanovska, L. (2005). Bioavailability of isoflavone phytoestrogens in postmenopausal women consuming soya milk fermented with probiotic bifidobacteria. *Br J Nutr,* 93, 867–877.

Tuberoso, C.I.G., Kowalczyk, A., Sarritzu, E., Cabras, P. (2007). Determination of antioxidant compounds and antioxidant activity in commercial oilseeds for food use. *Food Chemistry*, 103, 1494-1501.

Umpierrez, G.E., Spirgelman, R., Zhao, V., Smiley, D.D., Pinzon, I., Griffith, D.P., Peng, L., Morris, T., Luo, M., Garcia, H., Thomas, C., Newton, C.A. & Ziegler, T.R. (2012). A double-blind, randomized clinical trial comparing soybean oil– based versus olive oil–based lipid emulsions in adult medical–surgical intensive care unit patients requiring parenteral nutrition. *Crit C Med*, 40, 1792-1798.

Upmalis, D.H., Lobo, R., Bradley, L., Warren, M., Cone, F.L. & Lamia, C.A. (2000). Vasomotor symptom relief by soy isoflavone extract tablets in postmenopausal women: a multicenter, double-blind, randomized, placebo-controlled study. *Menopause,* 7, 236-242.

Valachovicova, T., Slivova, V. & Sliva, D. (2004). Cellular and physiological effects of soy flavonoids. *Mini Rev Med Chem*, 4, 881–887.

Vaz, J.S., Kaca, G., Nardib, A.E., & Hibbelnc, J.R. (2014). Omega-6 fatty acids and greater likelihood of suicide risk and major depression in early pregnancy. *Journal of Affective Disorders*, 152, 76-82.

Weisberg, S.P., McCann, D., Desai, M., Rosenbaum, M., Leibel, R.L. & Ferrante, A.W. Jr. (2003). Obesity is associated with macrophage accumulation in adipose tissue. *J Clin Invest,* 112, 1796-1808.

Wendlinger, C., Hammann, S. & Vetter, W. (2014). Various concentrations of erucic acid in mustard oil and mustard. *Food. Chem.*, 153, 393-397.

Whittinghill, J. & Welsby, D. (2010). Use of SDA soybean oil in bakery applications. *Lipid Technol*, 22, 203–205.

Wilson, R.F. (2004). Seed composition. In H. Boerma, J.E., Specht, (Eds.), Soybeans: improvement, production, and uses (3rd ed.), ASA, CSSA and SSSA: Madison, pp. 621–668.

Yan, L. & Spitznagel, E. (2005). A meta-analysis of soy food and breast cancer risk in women. *Int J Cancer Prev*, 1, 281–293.

Yaqoob, P. & Shaikh, S.R. (2010). The nutritional and clinical significance of lipid rafts. *Curr Opin Clin Nutr Metab Care*,13, 156–66.

Yettella, R.R., Henbest, B. & Proctor, A. (2011). Effect of antioxidants on soy oil conjugated linoleic acid production and its oxidative stability. *J Agric Food Chem,* 59, 7377-7384.

Yi, M.A., Son, H.M., Lee, J.S., Kwon, C.S., Lim, J.K., Yeo, Y.K., Park, Y.S. & Kim, J.S. (2002). Regulation of male sex hormone levels by soy isoflavones in rats. *Nutr Cancer*, 42, 206-210.

Zhang, P., Burton, J.W., Upchurch, R.G., Whittle, E., Shanklin, J., Dewey, R.E. (2008). Mutations in a D9-stearoyl-ACP-desaturase gene are associated with enhanced stearic acid levels in soybean seeds. *Crop Sci,* 48, 2305–2313.

Zhang, X., Shu, X.O., Li, H., Yang, G., Li, Q., Gao, Y.T. & Zheng, W. (2005) Prospective cohort study of soy food consumption and risk of bone fracture among postmenopausal women. *Arch Intern Med*, 165, 1890-1895.

In: Seed Oil
Editor: Alexis Varnham

ISBN: 978-1-63463-056-6

Chapter 2

CHARACTERIZATION OF ARGENTINEAN CHIA SEED OIL OBTAINED BY DIFFERENT PROCESSES: A MULTIVARIATE STUDY

***Vanesa Y. Ixtaina*[1], *Susana M. Nolasco*[2] *and Mabel C. Tomás*[1]**
[1]Centro de Investigación y Desarrollo en Criotecnología de Alimentos (CIDCA), (CONICET La Plata – UNLP), La Plata, Buenos Aires, Argentina
[2]Grupo de Investigaciones TECSE. Departamento de Ingeniería Química. Facultad de Ingeniería, UNCPBA, Olavarría, Buenos Aires, Argentina

ABSTRACT

Chia (*Salvia hispanica* L.) seed oil is a very interesting source with regard to provide a good equilibrium between two essential fatty acids (FAs) (linoleic and α-linolenic acid). Currently, chia seed oil is not widely used commercially even though its characteristics are well-suited for industrial applications, and contribute to healthy human diets. One of the main objectives of chia oil production involves the appropriate selection of the extraction process. The yield and the quality of oil are very important to determine the feasibility of commercial production. Chia seed oil was obtained by different extraction processes, some of them commonly used by the oil industry (solid-liquid extraction and cold pressing) or by alternative technologies with supercritical CO_2 (SC - CO_2). The aim of this work was to analyze the oil yield, the fatty acid composition, the total tocopherol and polyphenolic compounds content and the oxidative stability of chia seed oils obtained by solvent, pressing and CO_2 supercritical extraction (CO_2-SE) by a multivariate statistical method. The highest oil yield was 0.34 g/g seed (d.b.) obtained by solvent extraction (hexane). It was also possible to achieve similar values by adjusting the operating conditions (pressure, temperature and time of extraction) of the SC-CO_2 process. However, the oil yield reached by pressing was about 30% lower than those obtained by solvent (hexane) and SC-CO_2. The fatty acid composition of oils was similar for the different processes, highlighting the α-linolenic (~65%) and linoleic (~20%) acids content and a low level of saturated acids (~9%). Furthermore, the presence of a moderate amount of bioactive compounds such as tocopherols and polyphenols, was recorded. Multivariate analysis showed that the first three principal components described about 92% of the variance. The features that differentiate the oils obtained by conventional processes from those extracted by CO_2-SE were the presence of larger amounts of oleic and stearic acids, tocopherols and oxidative stability in the former, and the increased quantities of palmitic and linoleic (C18:2) acids and total polyphenol compounds in the latter.

INTRODUCTION

The importance of fats for humans, animals and plants lies in their high content of energy. In addition, fats allow humans and animals to consume fat-soluble vitamins and provide them with essential fatty acids (FAs), which are indispensable because their bodies are unable to synthesize themselves (Bockisch, 1998).

Vegetable oils are used for many food and industrial purposes. Although a wide variety of sources of vegetable oils, global consumption is dominated by palm, soybean, rapeseed and sunflower oils. In recent years there has been development of underexploited promising plant species as a source of dietary or specialty oils. Many of them contain significant quantities of oils and/or a high proportion of nutritionally, medicinally or industrially desirable FAs.

Chia seeds (*Salvia hispanica* L.) have a long history in the plant-human interaction. In pre-Columbian Mesoamerica the crop species was a major commodity and its seeds were valued for food, medicine and oil (Ayerza, 1995). Today, *S. hispanica* is mostly grown in Mexico, Bolivia, Argentina, Ecuador and Guatemala and it has been demonstrated that the species has great potential as a future crop plant (Coates and Ayerza, 1996). Chia seeds contain about 32-39% of oil which presents the greatest α-linolenic acid (C 18:3) content known up today (61-70%). Furthermore, the seeds have natural antioxidants which contribute to the oil preservation, inhibiting or delaying the development of the off-flavors that reduce the acceptability by the consumers (Ixtaina et al., 2008). Nowadays, chia seed oil is receiving increased attention, since it can improve human nutrition by providing a natural, plant-based source of ω-3 FA and antioxidants.

One of the main objectives of oil production is the proper selection of the extraction method. The extraction yield and the quality of the oil are very important to determine the feasibility of commercial production.

Liquid-solid extraction, mainly using hexane as solvent, is one of the more traditional processes employed in the production of seed oils. The solvent extraction principle is based on the fact that a component (solute) is distributed between two phases according to an equilibrium determined by the nature of the component and the two phases (Bockisch, 1998). In order to facilitate the extraction process is necessary to reduce the size of the seed or grain or even broken by the rolling process (Prámparo et al., 2003). The object of the extraction process is to reduce oil content in the flake to the lowest possible level with a minimum use of solvent (Milligan and Tandy, 1984). The application of a heat treatment before or during extraction causes cell breakage emulsion, reduces the oil viscosity and fluidity to facilitate movement and lowers the surface tension of the oil. However, such treatment may adversely affect the quality of the oil, increasing its oxidation parameters.

Furthermore, organic solvents such as hexane pose safety risks and health and environmental hazards and its replacement is being sought by the oil industry.

In recent years, there is an increased interest in the production of oils by cold-pressing technologies. For obtaining nontraditional vegetable oils this process provides an easy way to get oil from small seed lots (Wiesenborn et al., 2001; Zheng et al., 2003).

Although oil yields obtained by pressing are lower than those achieved using solid-liquid extraction. This technology is suitable for materials with high oil content because requires less expensive equipment and involves safe operation and lower risk for the environment. The press extraction principle is based on each particle retains the oil inside and the objective of pressing is to make that the oil migrates from the system to the outside. The application of an external force during the pressing produces a series of changes (deformations) both microscopically (cells) as macroscopic (Mattea, 1999).

Nowadays, the extraction of vegetable oils with solvents under supercritical conditions has been proposed as an alternative to replace conventional process (pressing, solvent extraction). This process ensures the absence of traces of solvent in the extracted oil, and allows more efficiently preserving its chemical and organoleptic properties (Norulaini et al., 2009).

The extraction principle is based on bringing the fluid to a specific supercritical state to extract a particular solute. Thus, the material to be subjected to the process is exposed under conditions of time, temperature and pressure controlled, allowing the dissolution of the solute of interest in the supercritical fluid. The dissolved solute is then separated from the supercritical fluid by reducing the pressure (Nielsen, 1998). The properties of supercritical fluids can be varied by changing the temperature and pressure. Thus, adjusting these parameters above the critical point can improve their ability to penetrate the structures of molecules and extract certain types of materials (Dunford et al., 2003). CO_2 is the most commonly supercritical fluid used for the extraction of food because it has a number of advantages: inexpensive, non-toxic, non-flammable, easily removed from extracts, high interpenetration in solid matrices. In processing terms, carbon dioxide has a low critical temperature and pressure (31.1°C and 73.8 atm, respectively), which make it the ideal solvent for natural products, since they do not suffer thermal degradation reactions during the process (Follegatti-Romero et al., 2009).

Supercritical extraction using CO_2 of various oil seeds, such as soybean, safflower, cottonseed, canola, millet bran, rice and other oils rich in ω-3 FA (fish and flaxseed oil) has been reported (Stahl et al., 1980; Friedrich and Pryde, 1984; Bozan and Temelli, 2002).

In this work, the oil yield, the fatty acid composition, the total tocopherol and polyphenolic compounds content and the oxidative stability of chia seed oils obtained by solvent, pressing and CO_2 supercritical extraction (CO_2-SE) were evaluated and the data obtained further analyzed by a multivariate statistical method verifying its ability to distinguish groups of oils.

Materials and Methods

Seeds

Commercial chia seeds were purchased from Functional Products S.A., Argentina. They were manually cleaned, homogenized and packed in hermetic plastic vessels and stored at 5°C until further use.

Randomized samples (approximately 7-8% d.b moisture content), picked by a sample splitter, (CPASA, Centro Proveedor Agropecuario, Buenos Aires, Argentina) were used to obtain oils by the different processes. Oils were analyzed in duplicate or triplicate.

Liquid-Solid Extraction (Solvent Extraction)

The extraction was made from seed samples previously grinded using a coffee mill (Braun, Type 4041, Mexico) for 60 s. It was carried out using *n*-hexane in a Soxhlet apparatus by thermal cycles at 80ºC for 8 h, following the IUPAC Standard Method (IUPAC, 1992). The solvent was removed using a rotary vacuum evaporator at 40ºC (Büchi, Flawil, Switzerland), under nitrogen stream. The oil content was gravimetrically determined and expressed as weight percent on dry basis (%, d.b.).

Pressing

The moisture content of the seed was adjusted to 10% for increasing the oil yield and to avoid problems of choking during the pressing process. The extraction was carried out in one step at 25-30°C using a pilot scale Komet screw press (Model CA 59 G, IBG Monforts, Mönchengladbach, Germany).

The restriction dye and the screw speed were 5-mm and 20 rpm, respectively, which were selected from previous work. Running temperature was checked with a digital thermometer inserted into the restriction dye.

The oil content was gravimetrically determined and expressed as weight percentage on dry basis (%, d.b.).

Supercritical CO_2 Extraction (CO_2-SE)

The extraction was carried out on a pilot plant system (extractor volume 1.5 L) with a single step separation and solvent recycle capacity. Extraction experiments were done at two pressure (250 and 450 bar) and temperature levels (40 and 60°C) with a CO_2 mass flow rate of 8 kg/h which was measured with a mass flowmeter (Rheonik, Germany). The extracts were collected in the separator vessel at 60 bar and 40°C. In each experiment, about 500 g of ground chia seeds were used.

The end of the extraction was set when the difference between two consecutive measurements of oil extracted was ≤ 0.001 g oil/g dry seeds. For this reason, the total extraction time was different for each operative condition, as follow: 285, 423, 135 and 138 min for 40°C-250 bar, 60°C-250 bar, 40°C-450 bar, 60°C-450 bar, respectively.

Oil Storage

Oils obtained by the different processes were stored in dark vessels with a nitrogen atmosphere at 4°C until their use.

Oil Analytical Determinations

The fatty acid composition was determined as methyl esters: 100 mL oil plus 1 mL 10% KOH in methanol were heated for 45 min at 85°C. Non-saponifiable lipids were extracted with petroleum ether (b.p. 30-40°C). After acidification with HCl, saponified FAs were extracted from the methanolic phase with petroleum ether. Fatty acids were methylated with 1 mL boron triflouride–methanol-complex (20% solution in methanol) (Merck) plus 1 mL methanol for 45 min at 60°C, and then extracted from the methanolic phase with petroleum ether. GC analysis: 1 mL hexane solution of fames was injected on column in GC (Hewlett Packard 6890) equipped with a capillary column Supelco 11090-02A Omegawax (30 m x 0.250 mm, i.d. 25 mm). The separation was carried out at 175-220°C (3°C/min) with helium as carrier (25.1 psi) and a FID detector at 260°C (Christie, 2003). The results were expressed as the relative percentage of each individual fatty acid (FA) presents in the sample. Oil tocopherol content was determined by normal phase HPLC using a Hewlett Packard chromatography system (HPLC Hewlett Packard 1050 Series, Waldbronn, Germany) equipped with a fluorescence detector Agilent 1100 Series (Agilent Technology, Palo Alto, CA, US) following the procedures described in IUPAC 2.432 (IUPAC, 1992) and AOCS Ce8-89 (AOCS, 1998). Total polyphenol content was analyzed by HPLC/APCI-MS according to Ixtaina et al., 2011a. These analyses were carried out with a Surveyor Plus Chromatograph coupled to a LTQ XL Linear Ion Trap (Thermo Fisher Scientific) The chromatographic separations were performed with a C_{18} 150mm x 2.1mm 335µm XTerra (Waters) and guard column C_{18} 4mm x 2mm (Phenomenex), All the assays were carried out by duplicate.

Oil oxidative stability was evaluated by the Rancimat (Mod 679, Metrohm) method, using 5 g oil sample warmed at 98°C with an air flow of 20 L/h. Oil stability was expressed in terms of induction time (h).

Statistical Analysis

A multivariate statistical analysis of the data set from the fatty acid composition and physicochemical characteristics of the oils obtained by the different processes was performed using principal component analysis (PCA). PCS reduces the number of variables according to their redundancy and finds the new components as linear combinations of the variables, with the first principal component having the largest variance, the second principal component with the second largest variance and so on. At the end of the procedure, the multidimensionality of the system is reduced to the first two or three principal components that retain most of the information of the data set (Flagella et al., 2002). Data were processed using the Statgraphics Centurion XV.II for Windows software (Statpoint Technologies, Warrenton, VA, US).

Results and Discussion

The oil yield of chia seeds extracted by the different processes is presented in Figure 1.

The highest oil yield was 0.34 g/g seed (d.b.) by solvent extraction (hexane). It was also possible to achieve similar values by adjusting the operating conditions (pressure, temperature and time of extraction) of the SC-CO_2 process. However, the oil yield reached by pressing was about 30% lower than those obtained by solvent (hexane) and CO_2-SE.

FA profiles are presented in Table 1. The variability observed in FA composition was within the normal range found in chia seed oil (Ayerza, 1995; AOCS, 1998; Ixtaina et al., 2010, 2011a).

α-linolenic acid was the main FA in chia seed oils, ranging from 64.5 to 65.6%. Linoleic acid was the second most prevalent FA (19.7-20.3%), followed by palmitic (6.2-6.7%) and oleic (5.0-5.5%) acids. The concentration of stearic acid was the lowest.

The obtained data suggest the potential value-added use of these seed oils as dietary sources of essential fatty acids.

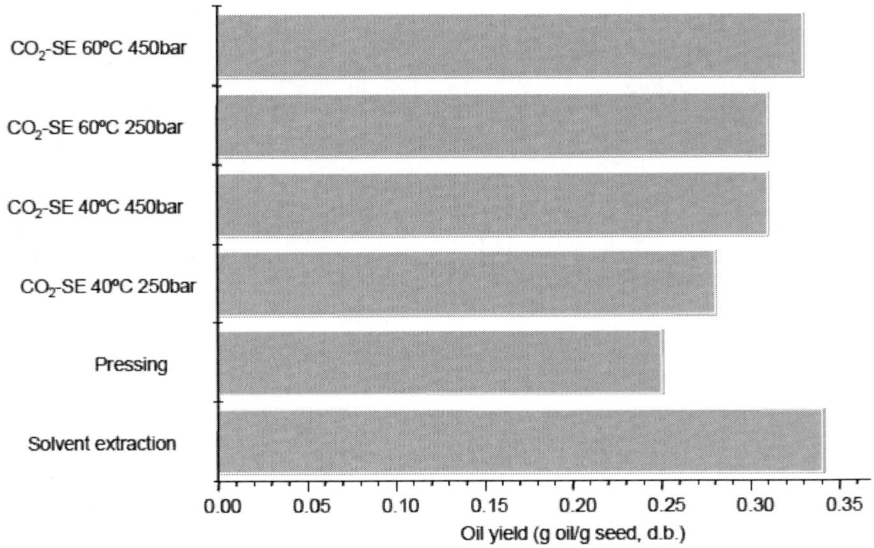

Figure 1. Yield of chia seed oil obtained by different processes. CO_2-SE, supercritical extraction using CO_2.

Table 1. Fatty acid composition (% of total FAs) determined by GC of chia seed oil obtaining by different processes

Extraction process		Fatty acid				
		Palmitic C16:0	Stearic C18:0	Oleic C18:1	Linoleic C18:2	α-linolenic C18:3
Solvent extraction		6.2	3.0	5.3	19.7	65.6
Pressing		6.6	3.1	5.4	20.3	64.5
CO_2-SE	40ºC - 250 bar	6.6	2.7	5.2	20.0	65.5
	60ºC - 250 bar	6.6	2.8	5.5	20.2	64.9
	40ºC - 450 bar	6.7	3.0	5.2	20.1	64.9
	60ºC - 450 bar	6.7	3.0	5.0	20.3	65.0

Mean values (n = 3).
CO_2-SE, supercritical extraction.

The ω-6/ ω-3 ratio of chia seed oils was about 0.3, being this value markedly lower than that of most vegetable oils, e.g. canola oil (2.2), olive oil (7.7), soybean oil (6.7) and walnut oil (5.0) (Belitz and Grosch, 1999).

Excessive amounts of ω -6 polyunsaturated fatty acids (PUFA) and a very high omega-6/omega-3 ratio, as is found in today's Western diets, promote the pathogenesis of many diseases, including cardiovascular disease, cancer, and inflammatory and autoimmune diseases, whereas increased levels of ω -3 PUFA (a low omega-6/omega-3 ratio) exert suppressive effects (Simopoulos, 2002). Therefore, the incorporation of chia seed oil into the diet would be very beneficial for human health.

The total amount of tocopherol showed a wide variation depending on the extraction process (Table 2). Oils extracted by CO_2-SE showed a low amount of these compounds. Tocopherols in chia oils were lower than those recorded in flaxseed (588.5 mg/kg), sunflower (634.4 mg/kg) and soybean (1797.6 mg/kg) oils (Tuberoso et al., 2007).

The total level of polyphenolic compounds was in the range of $5.30\ 10^{-5}$-$1.4x10^{-4}$ mol/kg (Table 2). These values are lower than those found in chia seeds (~$1.6x10^{-3}$ mol/kg). This fact is mainly related to the hydrophilic and polar nature of these compounds whose chemical structures therefore do not promote their oil solubility (Ixtaina et al. 2011b).

The accelerated stability test using Rancimat showed that chia oils have a low oxidative stability. The induction times ranged from 1.12 to 2.75 h, being the lowest values to those oils extracted by CO_2-SE. In spite of the presence of antioxidant compounds, the high content of PUFAs makes chia seed oil very instable. For this reason, the addition of natural antioxidants, such as green tea and rosemary extracts, ascorbyl palmitate and tocopherols, and also the storage conditions have been studied (Ixtaina et al., 2012).

Principal component analysis (PCA) was applied to the data set from the chia oils. The resulting score plot provided an overview of the oils in order to establish relationships between the oils extracted by different processes.

This plot reflected the differences among the oils and allowed to see the pattern of correlation between the variables.

Table 2. Total tocopherol and polyphenolic compounds, and oxidative stability (induction time) of chia seed oil obtaining by different processes

Extraction process		Total tocopherol (mg/kg)	Total Polyphenolic compounds (mol/kg)	Induction time (h)
Solvent extraction		295.3	9.15×10^{-5}	2.37
Pressing		238.4	8.90×10^{-5}	2.75
CO_2-SE	40°C - 250 bar	64.3	7.25×10^{-5}	1.12
	60°C - 250 bar	36.8	6.50×10^{-5}	1.22
	40°C - 450 bar	95.7	5.30×10^{-5}	1.60
	60°C - 450 bar	77.6	1.41×10^{-4}	1.53

CO_2-SE, supercritical extraction.

The variables found in a similar direction and far from the origin were positively correlated.

PC 1, 2 and 3 explain 43.5, 31.6 and 16.6% of the variability respectively, describing a total of about 92% of the variance. As can be seen from the Figure 2, PC 1 allowed separating the oils obtained by conventional process (pressing, solvent extraction) of those extracted by SC-CO_2, whereas PC2 clearly distinguished solvent extraction from pressing.

Thus, taking into account PC1, oils obtained by solvent and pressing were associated with high content of oleic (C18:1) and stearic acids (C18:0), tocopherols and oxidative stability. On the other hand, oils extracted by supercritical CO_2 (SC-CO_2) were related to high content of palmitic (C16:0) and linoleic (C18:2) acids and total polyphenol compounds.

Regarding PC2, the extraction process using hexane was associated with a high oil yield and α-linolenic (C18:3) acid content, whereas oils obtained by pressing were related to high levels of stearic, oleic and linoleic acids and high oxidative stability.

The PC3 clustered the oils obtained by CO_2-SE according to the operative condition. Thus, oils obtained at 250 bar (40-60°C) were related to a high content of oleic acid, whereas that extracted at 450 bar and 60°C was associated with high oil yield, stearic acid and total polyphenolic compounds.

The association between the variables showed that the oxidative stability is related mainly to the total tocopherol content, indicating the importance of these natural compounds as antioxidants. The total amount of polyphenols was inversely associated with the oxidative stability.

CONCLUSION

The application of a multivariate analysis to the chemical data showed that the pattern of variation for the FA, antioxidants compounds and oxidative stability could be visualized by the loading plot obtained by PCA.

The features that differentiated the oils obtained by conventional processes from those extracted by CO_2-SE were the presence of larger amounts of oleic and stearic acids, tocopherols and oxidative stability in the former, and the increased quantities of palmitic and linoleic (C18:2) acids and total polyphenol compounds in the latter.

The fatty acid composition and the presence of minor compounds confer the beneficial qualities on chia oil from the nutritional point of view, being of interest their potential application in the food industry.

The different extraction processes are associated to a greater or lesser extent with the different variables studied.

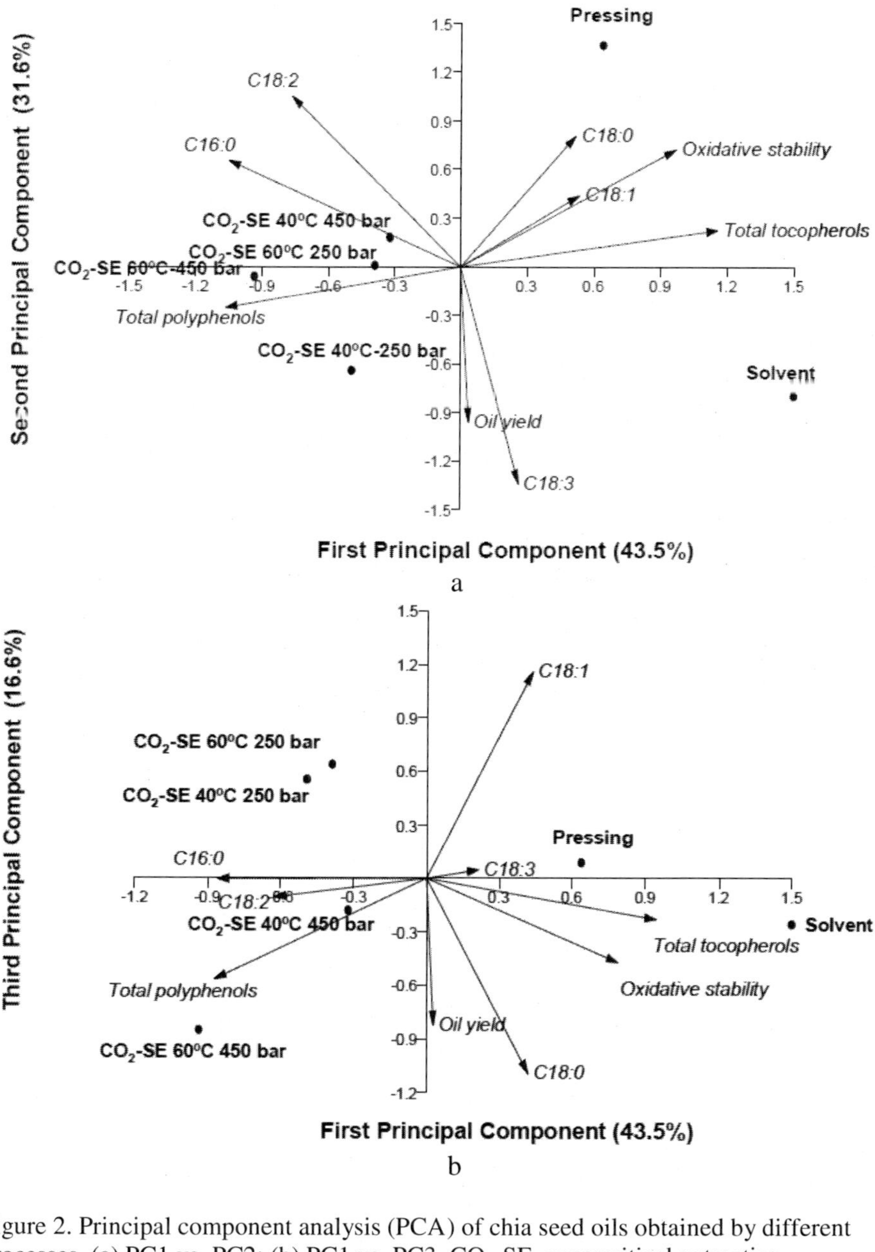

Figure 2. Principal component analysis (PCA) of chia seed oils obtained by different processes. (a) PC1 vs. PC2; (b) PC1 vs. PC3. CO_2-SE, supercritical extraction.

ACKNOWLEDGMENTS

This work was supported by grants from Universidad Nacional de La Plata (UNLP) (11/X610), PIP 1735 CONICET. The authors wish to thank Carmen Mateo, Margarita García and Viviana Spotorno for their technical support.

Author S.M. Nolasco is a Scientific and Technological Researcher and Professor at the Facultad de Ingeniería de la Universidad Nacional del Centro de la Provincia de Buenos Aires (UNCPBA); V.Y. Ixtaina and M.C. Tomás are members of the career of Scientific and Technological Researcher of the CONICET, Argentina.

REFERENCES

AOCS (1998) *Official Methods and Recommended Practices of the AOCS*, 5th ed., AOCS Press, Champaign.

Ayerza, R. (Jr) (1995). Oil Content and Fatty Acid Composition of Chia (*Salvia hispanica* L.) from Five Northwestern Locations in Argentina. *J. Am. Oil Chem. Soc.* 72: 1079-1081.

Belitz, H. D., Grosch, W. (1999). *Food Chemistry*, 2nd ed. Springer-Verlag, Berlin, Germany.

Bockisch, M. (1998). Extraction of vegetable oils. In: *Fats and oils handbook.* AOCS Press, Champaign, US.

Bozan, B., Temelli, F. (2002). Supercritical CO_2 extraction of flaxseed. *J. Am. Oil Chem. Soc.* 79:231-235.

Christie, W. W. (2003). *Lipid analysis: Isolation, separation, identification, and structural analysis of lipids*, 3rd edn. Bridgwater, England: Oily Press.

Coates, W., Ayerza, R. (Jr). (1996). Production Potential of Chia in Northwestern Argentina. *Ind. Crops Prod.* 5: 229-233.

Dunford, N. T., Teel, J. A., King, J. W. (2003). A continuous counter current supercritical fluid deacidification process for phytosterol ester fortification in rice bran oil. *Food Res. Int.* 36: 175-181.

Flagella, Z., Rotunno, T., Tarantino, E., Di Caterina, R., De Caro, A. (2002). Changes in seed yield and oil fatty acid composition of high oleic sunflower (*Helianthus annuus* L.) hybrids in relation to the sowing dateand the wáter regime. *Eur. J. Agron.* 17: 221-230.

Follegatti-Romero, L., Piantino, C., Grimaldi, R., Cabral, F. (2009) Supercritical CO_2 extraction of omega-3 rich oil from Sacha inchi (*Plukenetia volubilis* L.) seeds. *J. Supercrit. Fluid* 49: 323-329.

Friedrich, J. P., Pryde, E. H. (1984) Supercritical CO_2 extraction of lipid-bearing materials and characterization of the products. *J. Am. Oil Chem. Soc.* 61:223-228.

IUPAC (1992). *Standard methods for the analysis of oils, fats and derivates* (7th ed.). Eds. Paquot, C., Hautffenne, A. International Union of Pure and Applied Chemistry, Blackwell Scientific Publications Inc., Oxford.

Ixtaina, V. Y., Nolasco, S. M., Tomás, M. C. (2008). Physical properties of chía (*Salvia hispanica* L.) seeds. *Ind. Crops Prod.* 28: 286-293.

Ixtaina, V. Y., Vega, A., Nolasco, S. M., Tomás, M. C., Gimeno, M., Bárzana, E., Tecante, A. (2010). Supercritical carbon dioxide extraction of oil from Mexican chia seed (*Salvia hispanica* L.). Characterization and process optimization. *J. Supercrit. Fluids*. 55 (1): 192-199.

Ixtaina, V. Y., Martínez, M. L., Spotorno, V., Mateo, C. M., Maestri, D. M., Diehl, B. W. K., Nolasco, S. M., Tomás, M. C. (2011a) Characterization of chia seed oils obtained by pressing and solvent extraction. *J. Food Comp. Anal.* 24: 166-174.

Ixtaina, V. Y., Mattea, F., Cardarelli, D., Mattea, M., Nolasco, S. M., Tomás, M. C. (2011b). Supercritical carbon dioxide extraction and characterization of Argentinean chia seed oil. *J. Am. Oil Chem. Soc.* 88 (2): 289-298.

Ixtaina, V. Y., Nolasco, S. M., Tomás, M. C. (2012). Oxidative stability of chia (*Salvia hispanica* L.) seed oil: effect of antioxidants and storage conditions, *c* (6), 1077-1090.

Mattea, M. A. (1999). Fundamentos sobre el prensado de semillas oleaginosas. *Aceites Grasas*: 427-431.

Milligan, E. D., Tandy, D. C. (1984). Field evaluation of extraction performance. *J. Am. Oil Chem. Soc.* 61: 1383-1387.

Nielsen, S. S. (1998). Food analysis. In: *Crude fats analysis*. Purdue University, West Lafayette, US, pp. 203-214.

Norulaini, N. A. N., Setianto, W. B., Zaidul, I. S. M., Nawi, A. H., Azizi, C. Y. M., Mohd, O. A. K. (2009). Effects of supercritical carbon dioxide extraction parameters on virgin coconut oil yield and medium-chain triglyceride content. *Food Chem.* 116: 193-197.

Prámparo, M., Mattea, M., Gregory, S. (2003). Influencia del tipo de contacto sólido líquido en la extracción de aceites vegetales. *Aceites Grasas* 53: 592-597.

Simopoulos, A. P. (2002). The importance of the ratio of omega-6/omega-3 essential fatty acids. *Biom. Pharmacother.* 56: 365-379.

Stahl, E., Schutz, E., Mangold, H. K. (1980) Extraction of seed oils with liquid and supercritical CO_2. *J. Agr. Food Chem.* 28:1153-1157.

Tuberoso, C., Kowalczyk, A., Sarritzu, E., Cabras, P. (2007). Determination of antioxidant compounds and antioxidant activity in commercial oilseeds for food use. *Food Chem.* 103, 1494-1501.

Wiesenborn, D., Doddapaneni, R., Tostenson, K., Kangas, N. (2001). Cooking indices to predict screw-press performance for crambe seed. *J. Am. Oil Chem. Soc.* 78: 467-471.

Zheng, Y., Wiesenborn, D. P., Tostenson, K., Kangas, N. (2003). Screw pressing of whole and dehulled flaxseed for organic oil. *J. Am. Oil Chem. Soc.* 80: 1039-1045.

In: Seed Oil
Editor: Alexis Varnham

ISBN: 978-1-63463-056-6

Chapter 3

EFFECTS OF PRETREATMENTS ON THE YIELD AND QUALITY OF SUNFLOWER AND RAPESEED OILS

***M. B. Fernández*[1,2], *E. E. Pérez*[3] *and Susana M. Nolasco*[1]**
[1]TECSE - Facultad de Ingeniería - Universidad Nacional del Centro de la Provincia de Buenos Aires, Olavarría, Argentina
[2]CIFICEN (Universidad Nacional del Centro de la Provincia de Buenos Aires-CONICET), Tandil, Argentina
[3]PLAPIQUI (Universidad Nacional del Sur-CONICET), Bahía Blanca, Argentina

ABSTRACT

Rapeseed oil contains high amounts of bioactive compounds, such as polyphenols, phytosterols, tocopherols and other antioxidants, which play an important role in the prevention and treatment of some chronic diseases and improve immune function. In addition to its use as a food, this oilseed is also a viable option for the production of alternative fuels (biodiesel) due to its high oil content and yield per hectare, as well as the good quality of the extracted oil. Sunflower oil is used as a food and as an emollient in ointments and creams. Sunflower oil is essentially free of linolenic acid compared to soybean and rapeseed oils (3-10%). This provides some increased oxidative stability, but does not furnish valuable omega-3 acids that are necessary for health. Tocopherols are the main compounds with antioxidant properties present in sunflower seeds. In the oil extraction process, the seeds undergo a series of unit operations such as drying, storage, crushing, cleaning, flaking, conditioning, mechanical pressing and extrusion followed by solvent extraction. These processing stages may affect the quality and quantity of the oil extracted. First it is necessary to reduce the moisture content of the seeds for safe storage. The literature shows divergent data on the effect of the process temperature on the oil quality (measured in terms of the acidity value, peroxide index and tocopherol content) of different seeds. Conditioning of the seeds prior to extraction is required to make the oil inside the membranes more accessible to the solvent. Pretreatments such as crushing, hydrothermal treatments and the novel microwave technology are applied to seeds in order to modify or break their structure so as to facilitate the release of the oil. These pretreatments could also affect the release of other minor compounds, such as tocopherols. Another method used to make the release of the oil easier is by enzymatic degradation of the cell wall before and/or during extraction, but the release of bioactive compounds is also affected.

INTRODUCTION

The cultivated species that accumulate oil as reserve substances in their grains (seeds or fruits) are called oilseeds. Since ancient times, people have made use of the oils obtained from seeds and nuts as food or as raw material for producing biofuels (the latter mainly applies to rapeseed oil). These oils are used in raw and cooked food preparations and as the heat transfer medium in frying. Oils are essential nutrients, a source of calories and of fat-soluble vitamins, comprising about 40% of a person's daily calories [1, 2]. The world population growth raises questions about how demand for non-renewable resources of oil and food will be met. Projections by the United Nations estimate a world population of 16 billion by 2050. The worldwide oilseed production will face an increasing demand in the next thirty years due to a combination of factors, including higher consumption of edible oil, the development of the biofuel industry, and the need for green chemistry [3]. While there are many uses for industrial vegetable oils, total world production is only approximately 3% of that of edible oils [2]. The world's five major annual edible oilseeds are: soybean, cottonseed, rapeseed, sunflower and peanut [4, 5]. At present, the annual worldwide oil production is close to 135 Mt with palm, soybean and rapeseed oils representing 31%, 24% and 15% of total production, respectively [6]. The Russian Federation, Ukraine, Argentina, China, France, the United States of America, Eastern Europe and South Africa produce about 86% of the world's production ofsunflower seeds, while low erucic acid and low glucosinolate rapeseed consumption is higher in China, Canada, the European Union and Japan.

An important aspect of oilseed crops is that two products of economic value are mainly obtained: oil and meal. Most oilseeds have a higher oil concentration (e.g., sunflower, rapeseed) than protein (protein meal). Table 1 shows the average composition values of these oilseeds expressed as a percentage on dry basis (d.b.). Rapeseed and sunflower meal in general have a similar protein content, but lower than that of soybean meal, which is their main commercial end-use competitor.

Table 1. Chemical composition of major oil producing crops

Crop*	Oil content (% d.b.)	Protein content (% d.b.)	Fiber (% d. b.)	N-Free Extract (% d.b.)	Ash (% d.b.)
Cottonseed (delinted)	34	41	3	16	5
Peanut (kernel)	50	30	3	13	3
Rapeseed (dehulled)	44	25	5	20	5
Soybean (dehulled)	26	51	6	11	6
Sunflower (dehulled)	58	26	3	13	1

*Adapted from Bockisch M. [7].

Genetic Improvement of Rapeseed and Sunflower Seeds

The growing demand for oils not only for the food sector but also oleochemical industries is being met by the manipulation of the four major crops (soybean, palm, rapeseed and sunflower) using genetic engineering and the domestication of new oilseed crops. In the last 40 years, the yields of major oilseed crops has been increased considerably to satisfy demand and to allocate additional plots for the cultivation of species for industrial purposes. In turn, genetic manipulation makes it possible to obtain oils with fatty acid contents suitable for industrial use [3].

Species producing rapeseed oil and meal are the *Brassica* genus, of the family Cruciferae. Other members in the family include mustard seeds (for seasoning), vegetables such as cabbage, broccoli and others [8]. *Brassica napus* and *campestris* or *rapa* are the two most important oilseed species of the genus *Brassica* [9]. The origins of *Brassica* crops are a little uncertain. Several of these species appear to have been domesticated in a number of different places and dates as the plants became useful for local populations.

The history of rapeseed genetic improvement is related to the content of erucic acid (Figure 1a) in its oil and of glucosinolates in the meal (Figure 1b). In older rapeseed cultivars prior to 1973, erucic acid represented approximately half of the fatty acids. This acid is toxic, representing a serious health risk. Studies suggest that high levels of erucic acid fed to mice are associated with fatty deposits in the heart, skeletal muscle and adrenal glands of the rodents, also affecting their growth. Glucosinolates were also recognized as a problem in the rapeseed meal fed to poultry and ruminant animals. Glucosinolates interfere with the absorption of iodine from thyroid glands and further contribute to liver disease in poultry. Generally they have an adverse effect on the growth and weight gain of animals [8]. Genetic improvement has made it possible to reduce erucic acid and glucosinolate content in the oil and meal of rapeseeds, respectively. The terrm "double low" is used to describe varieties that are low in erucic acid glucosinolate and thus enable more widespread uses of this oilseed in food and animal feed. The CODEX Alimentarius established that low-erucic acid rapeseed oil must not contain more than 2% erucic acid (as % of total fatty acids).

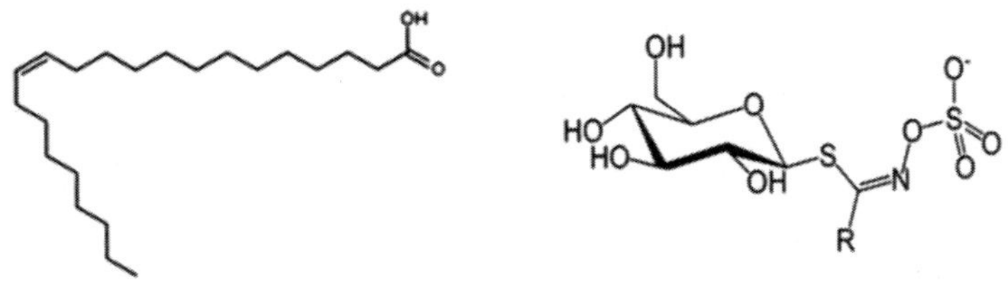

(a) (b)

Figure 1. Chemical structure of erucic acid (a) and a glucosinolate (b) molecule.

Regarding sunflower, seed companies have produced high-quality hybrid seeds by the identification of fertility-restorer genes. Open-pollinated cultivars were rapidly replaced by hybrids of higher yield, uniformity and disease resistance. Hybrid seeds are now widely used for cheap and efficient production throughout the world. Oil yields far exceed those under open pollination [10].There are three types of sunflower oil available in the market developed with standard breeding techniques: traditional or high linoleic, high oleic and mid-oleic sunflower oil. They differ in oleic levels and each one offers unique nutritional properties and industry requirements. These cultivars are caused by changes in the expression of the enzymes involved in fatty acid synthesis. This occurs by having fewer and reduced activity of the enzyme oleate desaturase [11, 12, 13], which is responsible for the transformation of the oleic acid into linoleic acid. Linoleic sunflower oil is the traditional sunflower oil and until recently it was the most common type of sunflower oil. It is predominantly (70%) polyunsaturated, with a light taste and is high in Vitamin E, whereas varieties of high oleic sunflower oil are very high in oleic, exceeding 70 %. The high monounsaturation makes high-oleic sunflower oil much less susceptible to oxidative degradation than traditional sunflower oil with high levels of polyunsaturation [14]. As a result, high-oleic oil is naturally stable and does not need to be hydrogenated. The mid-oleic sunflower cultivars produce oils with fatty acid concentrations of approximately 60-65%. In general, these cultivars were developed by crossing a traditional line and a high oleic, thus obtaining an intermediate oleic acid concentration between both lines. Sunflower cultivars with increased concentrations of saturated fatty acids, such as high stearic or high palmitic, have also been developed [15, 16]. These cultivars produce oils with lower fluidity, which is an important property for many industrial applications.

RAPESEED AND SUNFLOWER OILS CHARACTERISTICS

Low-erucic acid rapeseed oil has a low concentration of saturated fatty acids and contains linoleic (ω-6) and α-linolenic (ω-3) fatty acids in a ratio of approximately 2:1, and these characteristics make it one of the healthiest cooking oils. Alpha-linolenic acid belongs to the ω-3 family, and it is essential for normal human growth and development. FAO/WHO have recommended that the essential ω-6/ω-3 fatty acid balance in the diet should be between 5:1 and 10:1 [17]. Western diets are deficient in omega-3 fatty acids, and have excessive amounts of omega-6 fatty acids. Individuals who consume a ω-6:ω-3 ratio in excess of 10:1 should be encouraged to eat more ω-3 rich foods. Rapeseed oil also has significant levels of phytosterols, known inhibitors of cholesterol absorption. The importance of this oil not only resides in its nutritional value, but also in its physicochemical properties which make it a suitable raw material for the production of alternative fuels (high oil content and yield per hectare as well as good quality oil).

A comparison between the fatty acid composition of rapeseed and sunflower oil is shown in Figure 2. In addition to the difference in the amount of erucic acid, a great difference in oleic acid content between high erucic and low erucic rapeseed oil was observed. Comparing between species, unlike traditional sunflower oil, rapeseed oil contains significant levels of linolenic acid. Sunflower oil is essentially free of linolenic acid compared to rapeseed oils, which contain about 10% linolenic acid.

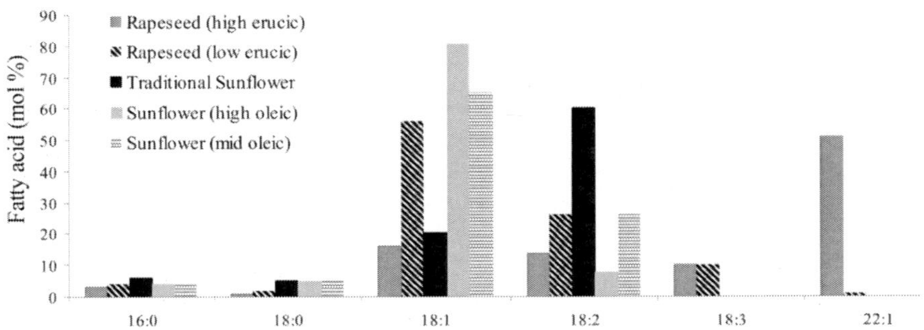

Figure 2. Fatty acid composition of sunflower and rapeseed oils. Adapted from refs. [18] and [19].

Table 2. Levels of sterols in crude vegetable oils, expressed as a percentage of total sterols

Sterol (%)	Rapeseed (low erucic)	Sunflower
Cholesterol	ND-1.3	ND-0.7
Brassicasterol	5.0-13.0	ND-0.3
Campesterol	24.7-38.6	5.0-13.0
Stigmasterol	0.2-1.0	4.5-13.0
β–Sitosterol	45.1-57.9	42.0-70
Δ5-Avenasterol	2.5-6.6	ND-6.9
Δ7-Avenasterol	ND-0.8	ND-9.0
Δ7-Stigmasterol	ND-1.3	6.5-24.0

ND – Non-detectable. Adapted from CODEX STAN 210-1999 (CODEX Alimentarius).

Regarding the minor compounds, Table 2 shows sterol composition expressed as a percentage of total sterol. In the case of rapeseed oil, total sterol content ranged 4500-11300 mg/kg oil, with β-sitosterol being the most abundant, followed by campesterol and brassicasterol.

On the other hand, sunflower oil showed a total sterol content of 1500-5200 mg/kg oil, with β-sitosterol being the most abundant, followed by Δ7-stigmasterol and stigmasterol. Total tocopherol content in rapeseed oil is in the 430-2680 mg/kg oil range. The main tocopherol is γ-tocopherol, followed by α-tocopherol. In the case of sunflower oil, total tocopherol content is in the 440-1520 mg/kg oil range, with α-tocopherol being the main compound (Table 3).

Table 3. Tocopherols profile in rapseed and sunflower oil (μg/g oil)

Compound (mg/kg)	Rapeseed (low erucic)	Sunflower
α-tocopherol	100-386	400-1090
β-tocopherol	ND-140	ND-52
γ-tocopherol	189-753	ND-34
δ-tocopherol	ND-22	ND-17

ND – Non-detectable. Adapted from CODEX STAN 210-1999 (CODEX Alimentarius).

Oil Processing

The existing procedures for lipid extraction from plant tissues usually involve several steps: pretreatment of the sample (which includes drying, size reduction, or hydrolysis), homogenization of the tissue in the presence of a solvent, separation of liquid (organic and aqueous) and solid phases, removal of nonlipid contaminants and removal of solvent, and drying of the extract. Considering extraction in the strict sense, three general types of processes are used to crush oilseeds: hard pressing, prepress solvent extraction and direct solvent extraction. The process of choice depends primarily upon the raw material, the amount of residual oil in the meal allowed, the amount of protein denaturation allowed, the amount of investment capital available and the local environmental laws concerning emissions of volatile organic compounds [20]. Solvent extraction is the most efficient method of extracting oil from the seed, generally leaving about 2% to 4% residual oil in the meal.

Figure 3 shows the typical schematic diagram for sunflower seed and rapeseed oil processing.

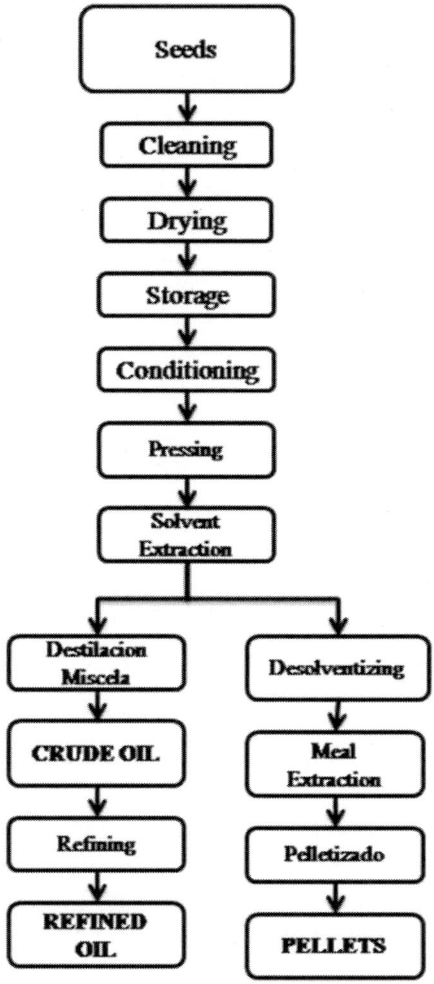

Figure 3. Schematic diagram of sunflower and rapeseed oil processing.

The conditioning depends on the oilseed. For example, dehulling of sunflower seeds is a required stage, but in the case of rapeseed, at present it is not a commercial process. Sunflower seed dehulling is necessary because the waxes sited in the hull tend to crystallize causing turbidity in the oil, affecting its processing and commercialization. The partial removal of the hull reduces the wax content and increases the protein content in the meal. Industrial dehulling is based on the impact of the grains at high speed, making the hull to break, and then it is separated by means of aspiration [21].

Pretreatments

The quality and stability of the oils and meals obtained during the extraction process are essential for commercialization and consumer acceptance. These properties depend mainly on the quality of the raw materials, harvesting and storage conditions, treatment of the oilseed before extraction, extraction method used and processing conditions, as well as the presence of some minor components.

a. Drying

In order to guarantee good storage or to condition the seeds before processing, the seeds should have the appropriate moisture content. If necessary, the seeds will have to be dried to the correct moisture content. In the drying process, variables such as temperature, time, characteristics of the dryer, among others affect the oil quality. Laoretani et al. (2014) did not find differences in acidity value, peroxide index and fatty acid profile when they analyzed the drying process of rapeseeds at 35 and 100 ºC at 13.6 and 22.7% initial moisture content (d.b.). Sutherland and Ghaly found similar results for rapeseed and sunflower seeds dried at up to 80 °C [22]. Pathak et al. observed no changes in free fatty acid content when rapeseed was dried from 20% initial moisture to 8% at the 50-95 °C range, but they did find differences in the acidity content of the oil between the treated samples and the control sample [23]. Bax et al. predicted the deterioration of crude sunflower oil after seed drying and observed a decrease in quality (peroxide index and acidity value) with temperature [24].

Capitani et al. found that temperature and storage time generated a decrease in tocopherol content in the oil of wheat germen samples at 27 °C and 45 °C [25].

On the other hand, the drying temperature affected the tocopherol content of the oil depending on the initial moisture of the samples. Table 4 shows α and γ-tocopherol content of oil extracted from untreated rapeseeds and seeds dried at different conditions of initial moisture and temperature [26]. It is worth mentioning that β and δ-tocopherols were only present in traces in these oils.

When the moisture content of the sample was high (22.7%, d.b.), it was negatively affected by the drying temperature in the 35-100 °C range, except at the lowest temperature studied. However, at lower moisture levels, it was possible to apply a drying treatment of 100 °C for short periods of time (3 min) without significantly affecting these parameters.

Table 4. Tocopherol content of oils extracted from untreated rape seeds and seeds dried at different conditions of initial moisture and temperature

Temperature	Initial moisture content: 13.6% (d.b.)		Initial moisture content: 27.7% (d.b.)	
	α-Tocopherol	γ-Tocopherol	α-Tocopherol	γ-Tocopherol
Untreated sample	286 (0.3)[a]	413 (3.3)[a]	302 (6.0)[a]	398 (3.0)[a]
35	288 (9.3)[a]	416 (1.1)[a]	295 (5.9)[a]	406 (7.9)[a]
60	293 (8.9)[a]	413 (13.1)[a]	255 (5.5)[b]	368 (8.9)[b]
82	288 (3.5)[a]	413 (8.1)[a]	230 (7.9)[c]	358 (4.8)[b]
100	292 (3.9)[a]	414 (2.8)[a]	194 (2.8)[d]	262 (2.0)[c]

Different letters in the same column indicate significant differences (Tukey's Test, $p<0.05$). Standard deviation in parentheses. Adapted from Laoretani et al. [26).

b. Hydrothermal Pretreatment

Incorporating different pretreatments such as mechanical, hydrothermal or enzymatic treatments to oilseed processing generates a change in the cellular structure of the grains, improving the yield of the oil extraction process.

In the case of hydrothermal pretreatments, there are several studies in the literature that analyze their effect on different matrixes. Soral-Śmietana and Krupa investigated the changes in the microstructure and macrocomponents of white beans subjected to a mild hydrothermal treatment, and its effect on physical and chemical parameters [27]. They found that this process could affect the nutritional value of the seeds.

Vaporization has been applied to condition rapeseeds, evaluating its effect on the cell structure destruction, obtaining improved mechanical properties [28], thus favoring the pressing stage prior to solvent extraction at industrial level. Moreover, the application of hydrothermal pretreatments has also been proposed in order to obtain a more efficient rapeseed dehulling prior to oil extraction [29].

Mohammadzadeh et al. studied the effect of hydrothermal pretreatments on the dehulling efficiency and the quality of the oil extracted from rapeseeds [30]. They concluded that the extension in time of high moisture content in the seeds affected the oil quality by promoting the hydrolysis of triglycerides, thus generating free fatty acids and reducing the shell life of the oil.

The effect of hydrothermal pretreatments on the antioxidant activity of rapeseed oil was evaluated by Szydlowska-Czerniak et al. [31]. The incorporation of expanders (heated with steam) to rapeseed and sunflower processing plants allowed to improve the mechanical and seed extraction properties [8].

Fernández et al. studied different operating conditions for hydrothermal pretreatments and found that steaming broken seeds at 120 ºC for 5 min caused a significant improvement in oil yield (20%), with no significant effect on the peroxide index [32]. A negative effect was observed on acidity, but its value did not exceed that established by the trading rules of the Canadian Oilseed Processors Association (COPA).

The oil yield of hydrothermally pretreated sunflower and rapeseeds can be observed in Figure 4. For comparison purposes, oil yield data of untreated seeds were also included. In all cases, a marked increase in this parameter due to the treatment applied could be noted.

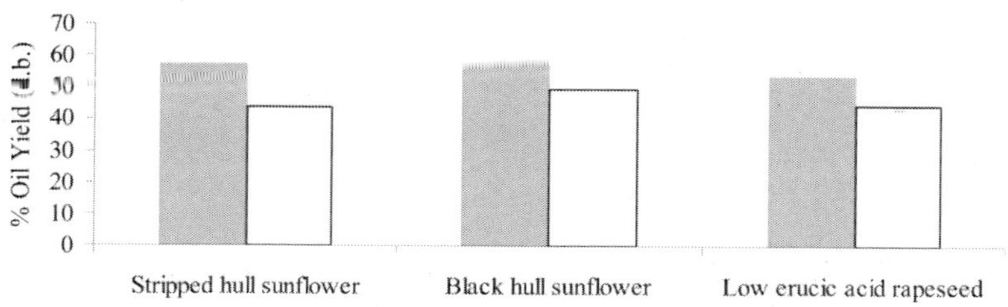

Figure 4. Oil yield from pretreated (■) and untreated (□) sunflower seeds and low erucic acid rapeseeds extracted in a Soxhlet apparatus [32, 33].

Figure 5 shows the kinetic data at 60 ºC for low-erucic acid rapeseed oil extracted from untreated seeds and seeds hydrothermally pretreated following the technique described by Zárate et al. [34]. For the hydrothermal pretreatment, the seeds were subjected to water steam in an autoclave. The pretreatment was carried out using broken seeds (particle size from 1.00 to 2.00 mm) at 393 K for 5 min. Then, they were dried to a moisture level of 6.5-7.4% d.b. The kinetics data were obtained in a batch device stirred with a magnetic agitator, and the solid/solvent ratio was 1:17. The results showed that the oil yield increased with the hydrothermal pretreatment and the processing time was reduced.

c. *Enzymatic Treatment*

The use of enzymes is another alternative method to facilitate the release of oil or other compounds of interest, as well as to find their beneficial effect on nutrition, and the quality and stability of the extracted products or byproducts [35]. It has been shown that the mixture of enzymes and complex multi-activity enzymes is more effective than the use of a single enzyme [36].

There are few research works on the production of edible sunflower oil using enzymes. The enzymatic action is not only affected by temperature, pH, the enzyme/substrate ratio and time of hydrolysis, but also by the treatment before and after the extraction. Enzymatic efficiency is not the same for the seeds or the meal, and it also varies depending on the type of hybrid.

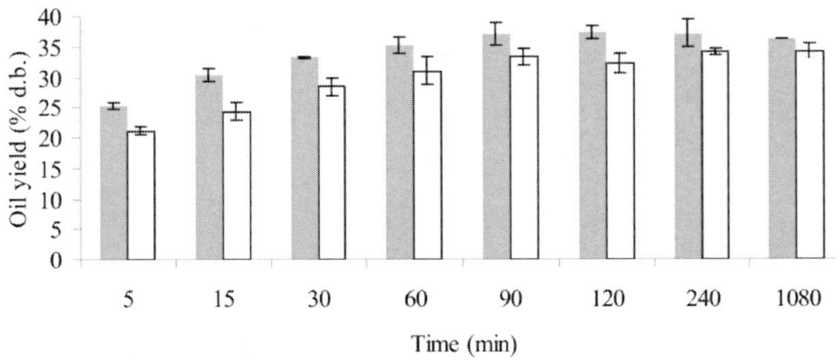

Figure 5. Oil yield from pretreated (■) and untreated (□) low-erucic acid rapeseeds vs extraction time. Extraction temperature: 60 °C.

Pérez et al. studied the pectinase-assisted oil extraction from two different sunflower genotypes [37]. Applying the enzymatic treatment effectively produced an increase in oil yield compared with the control samples (without enzyme) (Figure 6). The pectinase treatment was also highly effective in the extraction of tocopherols from a black hybrid sunflower, obtaining a 32.3% increase on average. Dominguez et al. found that an enzymatic treatment for seeds with high oil content, such as sunflower, not only enhanced the oil extractability in the pressing stage, but also made the residual oil in the cake more easily extractable by solvents [38].

d. Microwave Pretreatment

In the last decades there has been a growing demand for new pretreatment and/or extraction techniques that shorten the extraction times and reduce the consumption of organic solvents, while reducing pollution. These pretreatments include microwave radiation, ultrasound and enzymatic pretreatments. Microwave technology offers reduced processing times and energy savings because the energy is directly delivered to the material by molecular interaction with the electromagnetic field, so that the heat is generated throughout the volume of the material and can achieve rapid and uniform heating of relatively thicker materials [39]. The increase in oil yield for extraction by percolation (Soxhlet, hexane, 4 h) with microwave pretreatment can be observed in Figure 7. The oil yield increased by 19 and 10% for 100% and 80% microwave power, respectively [40].

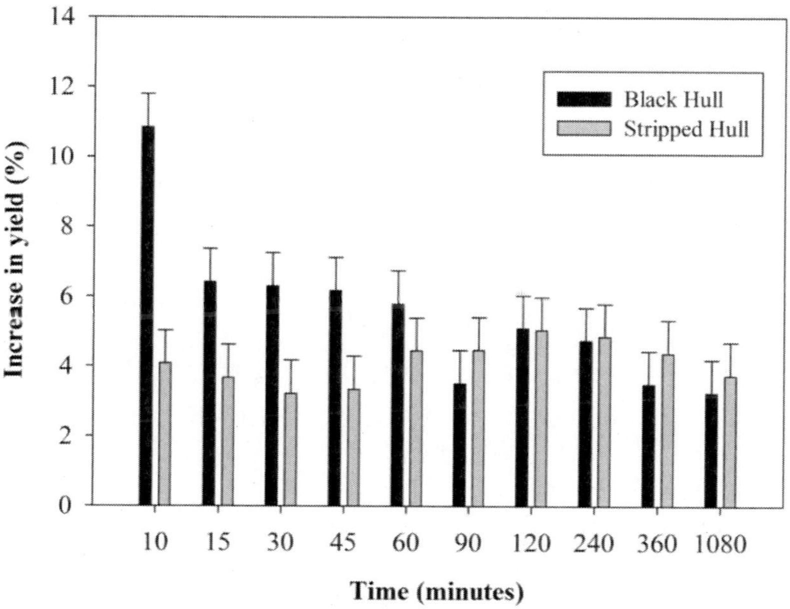

Figure 6. Percentage increase in oil yield over extraction time of stripped and black sunflower seeds treated with pectinase. Extraction temperature: 50 ºC.

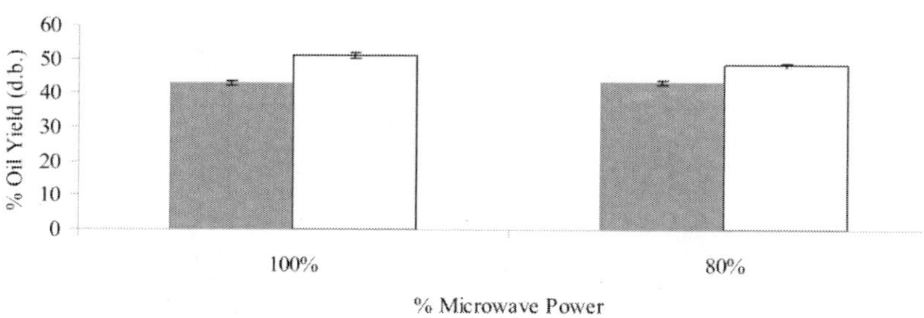

Figure 7. Oil yield from microwave pretreated (■) and untreated (□) rapeseeds. Conditions: initial moisture content 5.7% and 5.2% (d.b.), exposition time 5 and 4.1 min, at 100 % and 80 % of power, respectively.

CONCLUSION

The effects of different pretreatments on the yield and quality of sunflower and rapeseed oils were analyzed. In the drying process, operational variables affected the oil quality. Temperature affected the tocopherol content in oil at high initial moisture contents, but it had no effect at lower moisture levels. Hydrothermal, enzymatic and microwave pretreatments improved the release of oil.

REFERENCES

[1] Lawson, O. S.; Adefemi, O.; Ologunagba F. O.; Ojomo A. O. *J Eng Appl Sci.* 2010, 5, 51-55.

[2] Hassanein, M. M.; El-Shami, S. M.; Hassan El-Mallah, M. *Grasas y aceites* 2003, *54*, 343-349.

[3] Pérez, E.E.; Nolasco, S.M. In: *Advances in Fats and Oils Research*, Editor: M. C. Tomás, Transworld Research Network, Trivandrum, 2010, pp 49-69.

[4] Lusas, E. *Food Science and Technology* 2000, 297-362.

[5] McKevith, B. *Nutrition Bulletin* 2005, 30, 13-26.

[6] Jolivet, P.; Deruyffelaere, C.; Boulard, C.; Quinsac, A.; Savoire, R.; Nesi, N.; Chardot, T. *Industrial Crops and Products* 2013, 44, 549-557.

[7] Bockisch, M., *Fats and oils handbook*. AOCS press: 1998.

[8] Eskin, N.A.; Michael, M. D. B. E.; Prybulsky, R.; Malcolmson, L.J.; Scarth, R.; Mag, T.; Ward, K.; Adolph, D. *A&G* 2006, 4, 610-625.

[9] López Bellido, L. *Cultivos industriales*. Mundi Prensa Libros SA: 2003.

[10] Carbone, V; Carelli, A.A.; Carrín, M.E.; Cozzolino, R.; Crews, C.; Aparicio, R.; Ayadi, M.; Baumler, E.R.; Penci, M.C. In book: Mapping and comparing Oils Handbook, Chapter: Assessment of Chemical and Physical-Chemical Properties of Edible Oils. The Scientific Handbook, Publisher: Comunidad Económica Europea, Editors: Contract N 043083 Sixth framework programme. Priority 5. Food Quality and Safety Priority, Call 4C Specific Support Action Mapping and Comparing Oils (MAC–Oils) con la Comunidad Económica Europea, pp.127-198

[11] Garcés, R.; Mancha, M. *Phytochemistry* 1991, 30, 2127-2130.

[12] Kabbaj, A. A; Abbot, A.; Berville, A. In *Expression of stearate, oleate and linoleate desaturase genes in sunflower with normal and high-oleic*

contents, 14th International Sunflower Conference, Beijing, Shenyang, China,, 1996; Beijing, Shenyang, China,, 1996; p 6.

[13] Lagravere, T. C., L.; Lacombe, S.; Kleiber, D.; Berville, A. Y Dayde, J. In *Effects of temperature variations on fatty acids composition in oleic sunflower oil (Helianthus annuus L.) hybrids.*, 15th International Sunflower Conference., Toulouse, France, 2000; Toulouse, France, 2000; pp A-73.

[14] Dorrel, D. G.; Vick B.A., *Properties and Processing of Oilseed Sunflower In Sunflower Technology And Production* 1997.

[15] Osorio, J.; Fernández-Martínez, J.; Mancha, M.; Garcés, R. *Crop Science* 1995, 35, 739-742.

[16] Fernández-Martínez, J. M.; Mancha, M.; Osorio, J.; Garcés, R. *Euphytica* 1997, 97, (1), 113-116.

[17] FAO/OMS. Organización Mundial de la Salud. 1997. Grasas y Aceite en la Nutrición Humana. Consulta FAO/OMS de expertos., Roma. http://www.fao.org/docrep/v4700s/v4700s00.htm (Accessed August 2014)

[18] Przybylski, R.; Mag, T.; Eskin, N.A.M.; McDonald, B.E.; In Bailey's Industrial Oil And Fat Products. Sixth Edition.Volume 2. Cap 2 Canola oil. Edited by Fereidoon Shahidi.

[19] Grompone M.A.. In Bailey's Industrial Oil And Fat Products. Sixth Edition.Volume 2 Cap 14 Sunflower oil. Edited by Fereidoon Shahidi,

[20] Johnson, L. A. *Food Science and Technology-New York-Marcel Dekker-* 2002, 223-274.

[21] de Figueiredo, A.; Baümler, E.; Riccobene, I.; Nolasco, S. *Journal of Food Engineering* 2011, 102, 58-65.

[22] Sutherland, J.; Ghaly, T. *Journal of Stored Products Research* 1982, 18, 43-54.

[23] Pathak, P.; Agrawal, Y.; Singh, B. *Journal of the American Oil Chemists Society* 1991, 68, 580-582.

[24] Bax, M.; Gely, M.; Santalla, E. *Journal of the American Oil Chemists' Society* 2004, 81, 511-515.

[25] Capitani, M.; Mateo, C.; Nolasco, S. *Brazilian Journal of Chemical Engineering* 2011, 28, 243-250.

[26] Laoretani, D.; Fernández, M.; Crapiste, G.; Nolasco, S. *Antioxidants* 2014, 3, 190-199.

[27] Soral-Śmietana, M.; Krupa, U. *Czech J. Food Sci. Vol* 2005, 23, 74-83.

[28] Fornal, J.; Sadowska, J.; Winnicki, T. *Polish journal of food and nutrition sciences* 1996, 5, 25-36.

[29] Thakor, N. J.; Sokhansanj, S.; McGregor, I.; McCurdy, S. *Journal of the American Oil Chemists' Society* 1995, 72, 597-602.

[30] Mohamadzadeh, J.; Sadeghi-Mahoonak, A.; Yaghbani, M.; Aalami, M. *World J. Dairy Food Sci* 2009, 4, 14-18.

[31] Szydłowska-Czerniak, A.; Karlovits, G.; Sosna-Sárdi, Á.; Dianoczki, C.; Szłyk, E. *Journal of the American Oil Chemists' Society* 2009, 86, 817-825.

[32] Fernández, M. B.; Burnet, M. A.; Perez, E. E.; Crapiste, G. H.; Nolasco, S. M. *The Canadian Journal of Chemical Engineering* 2014, 92, 1239-1246.

[33] Burnet M.A., F. M. B., Crapiste G.H., Nolasco S.M., Perez E.E. In Procedings of *Congreso Argentino de Ingeniería Química 2010*, Mar del Plata, Argentina.

[34] Zárate, V.; Perez, E. E.; Crapiste, G. H.; Nolasco, S. M.; Fernández, M. B. *The Canadian Journal of Chemical Engineering* 2014, In Press.

[35] Ovando-Chacón, S. L., Waliszewski, K. N.. Universidad y ciencia 21, no. 42 (2005): 113-122. Rosenthal, A.; Pyle, D.; Niranjan, K. *Enzyme and Microbial Technology* 1996, 19, 402-420.

[36] Rosenthal, A.; Pyle, D.; Niranjan, K. *Enzyme and Microbial Technology* 1996, 19, 402-420.

[37] Perez, E. E.; Fernández, M. B.; Nolasco, S. M.; Crapiste, G. H. *Journal of Food engineering* 2013, 117, 393-398.

[38] Dominguez, H.; Sineiro, J.; Nuñez M. J.; Lema, J. M. *Food research internacional* 1995, 28, 537-545.

[39] Uquiche, E.; Jeréz, M.; Ortíz, J. *Innovative Food Science & Emerging Technologies* 2008, 9, 495-500.

[40] Ramos, L. Aplicación de tecnología emergente "microondas" como tratamiento previo a la extracción de aceite de canola. Tesis Licenciatura en Tecnlogía de los Alimentos. Universidad Nacional del Centro de la Provincia de Buenos Aires, Olavarría, Argentina, 2014.

In: Seed Oil
Editor: Alexis Varnham

ISBN: 978-1-63463-056-6

Chapter 4

THE USE OF MUCILAGE OBTAINED FROM VEGETABLE OIL SOURCES IN THE PREPARATION OF O/W EMULSIONS

Marianela I. Capitani[1,2], Susana M. Nolasco[2] and Mabel C. Tomás[1]

[1]Centro de Investigación y Desarrollo en Criotecnología de Alimentos (CIDCA), (CCT La Plata –CONICET) Facultad de Ciencias Exactas, La Plata, Buenos Aires, Argentina

[2]Grupo de Investigaciones TECSE. Departamento de Ingeniería Química. Facultad de Ingeniería, UNCPBA, Olavarría, Buenos Aires, Argentina

ABSTRACT

One of the factors that markedly affects the characteristics and stability of oil-in-water (O/W) emulsions is the presence of polysaccharides in the aqueous phase. O/W emulsions (20:80 wt/wt) were prepared with refined corn oil and dispersions with ≥0.75% mucilage (6.8 and 18.8% protein content) and 0.1% Tween 80, and they presented very good stability during storage for 120 days at 4±1°C (backscattering value 78%). The emulsions prepared with mucilage with lower protein (6.8%) and lipid content (0.9%) were more stable. The stability of O/W emulsions (40:60 wt/wt) prepared with canola oil and dispersions of mucilage extracted from locust bean and flax seeds was higher when the mucilage concentration was increased from 0.5 to 1.5% in the aqueous phase of the emulsion, whereas the emulsions formulated with mucilage from fenugreek seeds exhibited high stability (100%) during all the storage time (90 days, 25±1°C) and for all the mucilage concentrations tested (0.5-1.5%). It was also found that flax mucilage reduces the creaming phase in carrot juices, and helps to stabilize meat products by its interactions with the meat proteins. In general, polysaccharide dispersions increased the viscosity of the aqueous phase of the emulsions, limiting the mobility of the oil droplets in the dispersed phase to migrate, and therefore to flocculate or coalesce. Thus the physical stability of O/W emulsions against gravitational phase separation can be improved with the addition of chia mucilage, given its role as a thickening agent.

Keywords: Chia; flax; fenugreek; locust bean; mucilage; emulsion stability

INTRODUCTION

Emulsions are unstable systems consisting of two or more immiscible phases, a continuous and a dispersed phase (generally water and oil), where the dispersed phase forms small droplets inside the continuous phase. Emulsions can be classified according to the distribution of the phases. Systems wherein the oil droplets are dispersed in an aqueous continuous phase are called oil-in-water (O/W) emulsions (for example milk, mayonnaise, cream etc.), whereas systems wherein the water droplets are dispersed in an oil continuous phase are called water-in-oil (W/O) emulsions (for example butter and margarine) [1]. Multiple emulsions are also possible, including oil-in-water-in-oil (O/W/O) or water-in-oil-in-water (W/O/W) emulsions [2].

From the moment an emulsion is formed, a process of destabilization begins because the interfacial area of the droplets increases considerably and, therefore, the surface free energy of the system also increases. There exist different mechanisms that contribute simultaneously and synergistically with the destabilization, and that are generated by different physical phenomena related to the difference in density of the continuous and dispersed phase, colloidal interactions between the droplets, and the microstructure and viscoelasticity of the phases involved [1]. The main mechanisms of destabilization in emulsions are: *gravitational separation of phases*, caused by the upward (*creaming*, in O/W emulsions) or downward movement (*sedimentation,* in W/O emulsions) of the droplets depending on how much less or more dense they may be than the continuous phase, respectively [3]; *flocculation*, characterized by the aggregation of droplets, where each droplet maintains its identity without merging, generated by the effect of the thermal energy, gravity and the mechanical forces applied, with a decrease in the number of particles present in the emulsion; *coalescence*, process by which the droplets merge to form larger droplets, losing their identities, and it can take place by approximation, collision, deformation and rupture of the interfacial film; and *disproportionation* and *phase inversion*, corresponding to the change of an O/W emulsion into an W/O emulsion and vice versa, which can occur as a result of a large volume fraction of the dispersed phase, temperature or the application of mechanical forces [4, 1].

In order to obtain a better stability of an emulsified product, it should have the ability to resist change in its properties over time. In general, the stability of an emulsion is obtained by the presence of two types of ingredients: the "emulsifying agent" and the "stabilizing agent". The emulsifying agent is an amphiphile compound that tends to migrate and be adsorbed quickly in the oil-water interface, favoring the formation of the emulsion by reducing the interfacial tension and granting physical stability for a short period of time. Most emulsifiers used to stabilize food emulsions are classified into species of low molecular weight, such as lipids, phospholipids (lecithins), mono and diglycerides, polyoxyethylene sorbitan esters (Spans or Tweens), and species of high molecular weight, such as proteins and gums [5, 6, 7]. On the other hand, the stabilizing agent (mainly polysaccharides) provides physical stability to the emulsion for a long period of time by inhibiting the movement of the droplets of the dispersed phase due to an increase in viscosity of the aqueous phase [8, 9]. These substances also provide important characteristics to the product, for example a creamy mouthfeel and a high elastic limit [10]. Although most polysaccharides exhibit a stabilizing function, some can also have an emulsifying function. In order to fulfill their role as emulsifiers, polysaccharides must have the ability to act in the oil-water interface and thus facilitate the formation and stability of small droplets during and after emulsification. This function is also associated with the molecular structure, either by the non-polar character of the chemical groups bound to the structure of the hydrophobic polysaccharide or the presence of a protein component covalently or physically bound to the polysaccharide. Polysaccharides with an emulsifying action can also associate with the proteins adsorbed in the interface, favoring the stability of the emulsion [11]. Several polysaccharides such as gum arabic, modified starch, cellulose byproducts and some galactomannans such as guar gum, locust bean gum, and fenugreek gum present an emulsifying effect. However, polysaccharides must be added with caution because they can affect the colloidal interactions present in the emulsion and favor its destabilization by bridging flocculation or depletion flocculation, which could translate into an increase in creaming rate [12]. Thus, bridging flocculation occurs by a very weak interaction between polysaccharides and proteins when the concentration of the polysaccharide is not high enough to cover the whole surface of the droplets of the dispersed phase of the emulsion. The interfacial concentration of the hydrocolloid is the dominant factor controlling this mechanism, in addition to the properties of the polymer, such as molecular weight, degree of dissociation, molecular flexibility and surface hydrophobicity. On the other hand, depletion flocculation occurs when there are unabsorbed molecules of the polysaccharide given that the space between droplets is smaller than the thermodynamically more-stable hydrodynamic volume of the polysaccharide molecules. Therefore the polysaccharide is excluded from the space between the droplets, generating a layer around them where the polymer concentration is lower than in the bulk of the solution. Thus a local water concentration gradient is established, and consequently an osmotic pressure gradient that induces the flocculation [10] (Dickinson, 2003).

Mucilages Extracted from Seeds

Chia (*Salvia hispanica* L.) is an annual herbaceous plant native to Central America. In recent years, the potential use of this seed has been revalued given its quality and quantity of oil (33%), proteins, residual meal, dietary fiber and bioactive compounds [13, 14, 15]. Nowadays it is possible to find chia seeds in foods for human and animal consumption, being used in the production of bread, cookies, energy drinks and bars, dietary supplements, oil and products of animal origin (eggs, chicken, bovine meat, spicy sausage, jam, milk and cheese) enriched with ω-3 [16, 17, 18, 19, 20]. Many seeds contain reserve polysaccharides other than starch. Chia seeds contain mucilage (a complex polysaccharide of high molecular weight), which is secreted by the seed when it becomes wet [21]. The structural units of chia mucilage were described as a tetrasaccharide with a main chain consisting of units of (1→4)-β-D-xylopyranosyl-(1→4)-α-D-glucopyranosyl-(1→4)-β-D-xylopyranosyl with 4-O-methyl-α-D-glucuronic acid ramifications in the O-2 position of β-D-xylopyranosyl in the main chain. The ratio of β-D-xylose to α-D-glucose monosaccharides to 4-O-methyl-α-D-glucuronic acid is 2:1:1 [22, 23]. Chia mucilage presents high viscosity in water with possible beneficial metabolic effects. The intake of chia mucilage (alone or in combination with the seed) has shown to affect the metabolism of lipids by reducing the intestinal absorption of fatty acids, cholesterol and bile salts, increasing the loss of cholesterol through feces, and inhibiting endogenous cholesterol synthesis, slow digestion and absorption of nutrients. In addition, being part of the soluble dietary fiber, mucilage forms gels of high viscosity that produce gastric distension, feeling of satiety and slow stomach emptying, becoming a functional food [24]. The available data on its functional properties indicate that it is a polymer with thickening properties. In 1996, FAO described chia mucilage as a potential source of polysaccharides given its mucilaginous properties in aqueous solutions at low concentrations [25]. However, there is little information about its characteristics and potential applications as a stabilizing and emulsifying agent in the food industry.

Flax (*Linum usitatissimum* L.) is a crop mainly valued for the content, characteristics and uses of its oil (42% oil, with high content of polyunsaturated fatty acids) [26]. However, there is a growing interest in the use of this seed for human and animal consumption due to the presence of other bioactive compounds in its structure.

In this sense, flaxseeds contain approximately between 6.5 and 10.2% mucilage [27], which is generally obtained by aqueous extraction from the whole seed or meal. The yield and its characteristics depend on the extraction conditions, for example pH, seed:water ratio, time and temperature [28, 29]. Previous studies indicate that flax mucilage consists of an acid fraction, with a high content of galacturonic acid, and a neutral fraction with a low galacturonic acid content [29]. Some reports show that flax mucilage exhibits functional properties similar to those of gum arabic [28, 30]. Other studies report the capacity of this mucilage acting as emulsifier and stabilizing agent in the manufacture of chocolate [31], carrot juice [32, 33] and meat products [34].

Locust bean gum (LBG) is a galactomannan gum extracted from the seeds of the carob tree [35]. The seeds represent 10% of the pod and contain about 20 to 25% of oil [36], with mucilage yields between 26.7 and 40.0% [37, 38]. At present, the seeds of the carob tree are mainly used to make carob powder and to extract the mucilage, which is mainly used as thickening agent in the food industry due to its high viscosity in water, even at low concentrations, and for that reason it is used as a substitute for pectin, agar and other mucilaginous substances. This mucilage is also used as a medium to grow microorganisms, as food stabilizer and other applications in the textile, cosmetics and pharmaceutical industries [39, 40].

Fenugreek seeds (*Trigonella foenum-graecum*), with 7.24% oil, contain carbohydrate reserves in the endosperm that consist of 22.6% galactomannans [41]. The galactomannans present in fenugreek seeds are similar to those in guar gum and locust bean, and are used as thickening and stabilizing agents in the food industry [42]. Garti et al. [43] performed a study that shows the ability of fenugreek mucilage to stabilize O/W emulsions due to its surface activity associated with its residual protein content.

Preparation of O/W Emulsions with Seed Mucilages

Oil-in-water (O/W) emulsions (20:80 wt/wt) were prepared with refined corn oil, different dispersions of chia mucilage, 0.1% wt/wt of Tween 80 and 0.01% wt/wt of sodium azide to prevent bacterial growth. Both in the emulsions formulated with mucilage I (MI: 6.8% of protein and 0.9% of lipid) and mucilage II (MII: 18.8 protein and 3.2% lipid), the continuous phase was prepared so as to obtain a final composition of emulsions with mucilage concentrations of 0.25, 0.50, 0.75 and 1.00% wt/wt. The emulsions were prepared at room temperature in an Ultraturrax T-25 homogenizer and the resulting pre-emulsions were then treated with an ultrasonic processor. Then the samples were stored for 120 days at 4±1°C.

Emulsions prepared with dispersions of mucilage of locust bean, flax and fenugreek seeds were studied by Huang et al. [44]. O/W emulsions (40:60 wt/wt) were formulated with canola oil and different concentrations (0.5, 1.0 and 1.5%) of the different types of mucilage. These emulsions were stored for 90 days at ambient temperature (25°C).

The stability of the emulsions formulated with chia mucilage was analyzed as a function of storage time using a QuickScan vertical optical analyzer (Beckman Coulter, Fullerton, USA) as described previously by Pan et al. [45].

On the other hand, the initial stability of the emulsions prepared with dispersions of mucilage of locust bean, flax and fenugreek seeds was determined by a centrifugation test [46], and the stability as a function of time considering the variation in the height of the emulsion [44]. All the emulsions were also characterized by determining particle size distribution and viscosity.

CHARACTERISTICS OF THE EMULSIONS

The stability of the O/W emulsions prepared with 0.50% of different types of mucilage extracted from seeds of high lipid content, corresponding to the initial and final storage time, is presented in Figure 1.

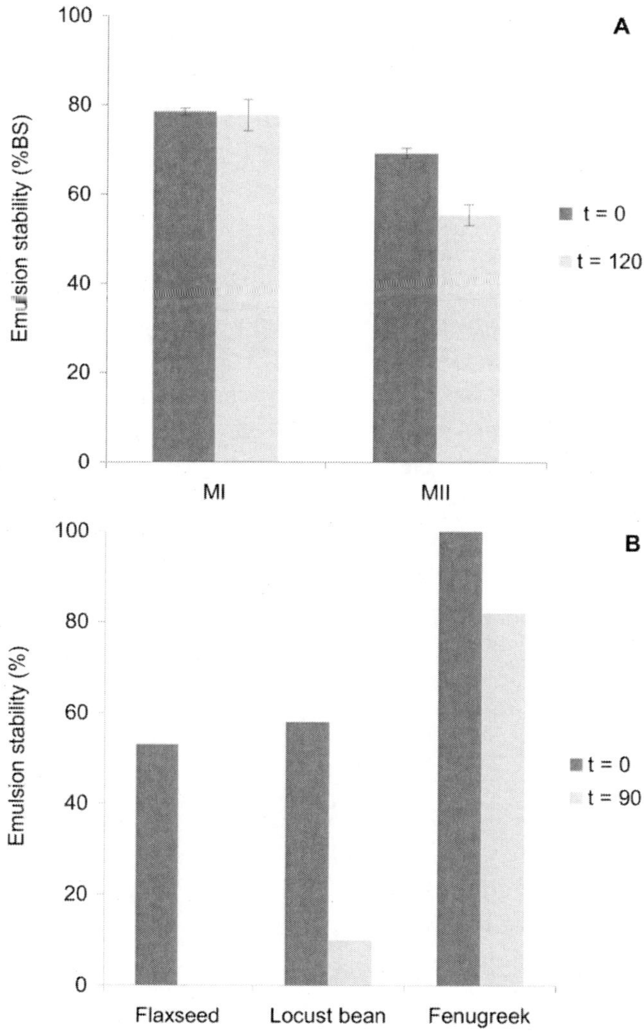

Figure 1. Stability of O/W emulsions formulated with 0.5% mucilage as a function of storage time (days). (**A**) 4±1°C with chia mucilage (MI and MII), (**B**) 25±1°C with flax, locust bean and fenugreek mucilages (adapted from Huang et al., [44]).

It can be observed that all the emulsions exhibit a high initial stability (>50%), which decreased with storage time, mainly for the emulsions formulated with mucilage of chia (MII), locust bean and flax seeds, by 28.7, 82.8 and 100% respectively, compared with the initial value. This behavior can be attributed to the fact that the presence of low concentrations of gums in an emulsion increases the rate of flocculation, coalescence and creaming of the emulsion [47].

A similar behavior was observed for O/W emulsions formulated with 0.2% of gum from *Lepidium perfoliatum* seeds [48]. Regarding the emulsions with chia mucilage, it is worth mentioning that those prepared with MI were more stable, and this could be ascribed to the lower protein content associated with this mucilage (6.8%).

Previous studies reported by Garti et al. [43] for fenugreek gum and by Wang et al. [49] for flax mucilage indicate that polysaccharides with certain protein content can either not affect or contribute to a better physical stability of the O/W emulsions. However, this behavior can change depending on the interactions between proteins and polysaccharides, because if the interactions are weak or repulsive, these systems may often present a phase separation at macroscopic level, and even at microscopic level [10].

On the other hand, the emulsions with higher mucilage concentration (1.00%) presented high stability along all the storage time for chia and fenugreek mucilage, whereas for the emulsions with locust bean and flax mucilage, stability decreased by 57.9 and 100%, respectively (Figure 2).

The high stability throughout storage time can be attributed to the increased viscosity of the continuous phase due to the high concentration of mucilage, thus reducing the mobility of the droplets of the dispersed phase [50]. It should be noted that the emulsion formulated with fenugreek mucilage was the only one that remained stable towards the end of storage time (90 days), according to the method used. Huang et al. [44] associated the high stability of this mucilage (63.4% by centrifugation) with its protein content (13.9%). However, they also consider that this does not seem to be an exclusive requirement (that the more stable emulsions are formulated with mucilage of high protein content), because emulsions formulated with flax mucilage (14.9% proteins) showed the lower stability (39.7% by centrifugation).

The Sauter mean diameters (related to the surface particle size distribution) of the emulsions formulated with 0.50% chia and fenugreek mucilage are presented in Table 1.

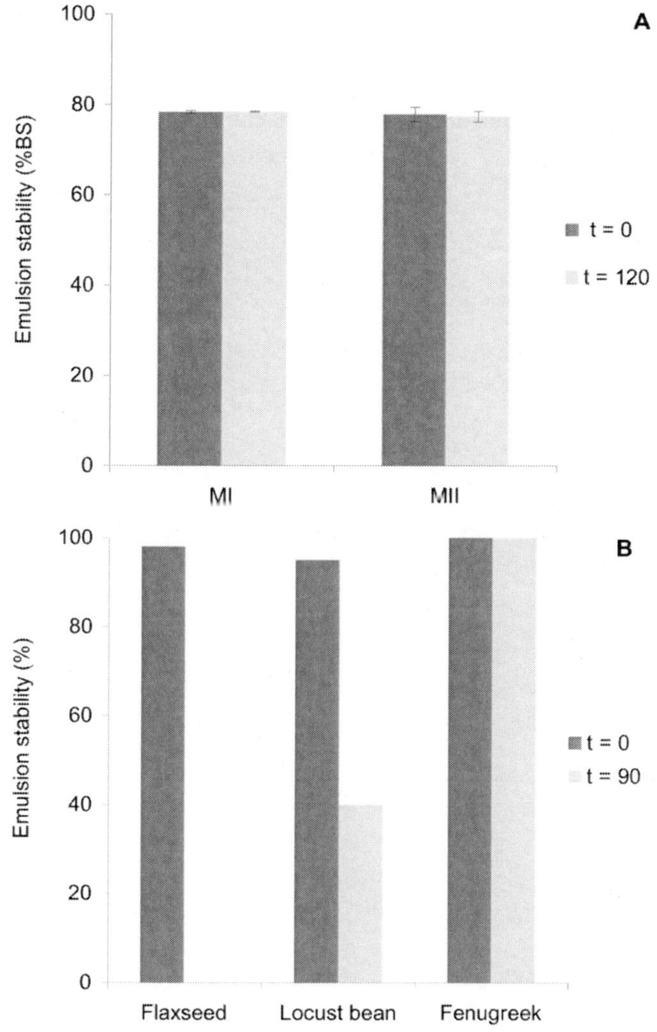

Figure 2. Stability of O/W emulsions formulated with 1.00% mucilage as a function of storage time (days). (**A**) 4±1°C with chia mucilage (MI and MII), (**B**) 25±1°C with flax, locust bean and fenugreek mucilages (adapted from Huang et al., [44]).

The emulsions with fenugreek exhibited the smaller droplet diameter. In the case of the emulsions with chia mucilage, they presented a larger particle size than those with fenugreek. This behavior could be associated with the higher viscosity of emulsions prepared with chia mucilage (Table 2), which could affect the homogenization process, hindering a complete rupture and distribution of the droplets of the emulsion. A similar behavior was reported for O/W emulsions formulated with xanthan gum [44].

Table 1. Sauter mean diameter (D[3,2]) of O/W emulsions formulated with 0.50% mucilage

Mucilage	D[3,2] (μm)
Chia – MI	2,57 ± 0,18
Chia – MII	3,10 ± 0,34
Fenugreek [a]	0,72
Xanthan [a]	2,61

[a] Huang et al., [44]

Table 2. Viscosity of O/W emulsions formulated with 0.50% mucilage

Mucilage	Viscosity (Pa s)
Chia – MI	0.0655 ± 0.02
Chia – MII	0.0275 ± 0.00
Fenugreek [a]	0.0143
Flaxseed [a]	0.0404
LBG [a]	0.1533
Xanthan [a]	0.9508

[a] Huang et al., [44].

CONCLUSION

The stability of O/W emulsions with added mucilage is influenced by the characteristics of the mucilage and its concentration. The addition of chia and funegreek mucilage at a 1.00% concentration produced emulsions that were stable throughout the storage period, whereas the stability of the emulsions with flaxseed and locust bean mucilage decreased considerably along storage time for the different concentrations studied. The type of mucilage used affects the size of the particles, as well as the viscosity of the emulsions.

REFERENCES

[1] McClements, D.J. Food Emulsions: principles, practice and techniques. Ed. CRC Press, New York (USA), 1999.

[2] Garti, N. *LWT - Food Sci Techn*, 1997, 30, 222-235.

[3] Walstra, P. Formation of emulsion. In: Encyclopedia of Emulsion Technology: Basic Theory. Ed. Becher P, Marcel Dekker Inc., New York (USA), 1983; pp 57-127.

[4] Campbell, I., Norton, I., Morley, W. *Neth Milk Dairy J,* 1996, 50, 167-182.

[5] Damodaran, S. Protein-Stabilized Foams and Emulsions. In: Damodaran S, Paraf A (Eds.), Food Proteins and their Applications, New York: Marcel Dekker, 1997; pp 57-110.

[6] Kamel, B.S. Emulsifiers. In: Smith J (Ed.), Food Additive User's Handbook. Glasgow: Blackie, 1991; pp 169-201.

[7] St. Angelo, A.J. A brief introduction to food emulsifiers. In: Charalambous G, Doxastakis G (Eds.), Food Emulsifiers: Chemistry, Technology, Functional Properties and Applications. Amsterdam: Elsevier, 1989; pp 1-8.

[8] Dickinson, E., Ma, J., Povey, M.J.W. *Food Hydrocolloid,* 1994, 8 (5), 481-497.

[9] Parker, A., Gunning, P.A., Ng, K., Robins, M.M. *Food Hydrocolloid,* 1995, 9(4), 333-342.

[10] Dickinson, E. *Food Hydrocolloid*, 2003, 17(1), 25-39.

[11] Hayati, I.N., Man, Y.B.C., Tan, C.P., Aini, I.N. *Food Hydrocolloid,* 2009, 23, 233-243.

[12] Tabilo-Munizaga, G., Barbosa-Cánovas, G.V. *J Food Eng,* 2005, 67(2), 147-156.

[13] Taga, M.S., Miller, E.E., Pratt, D.E. *J Am Oil Chem Soc*, 1984, *61*, 928-931.

[14] Ixtaina. V.Y., Martínez. M., Spotorno, V., Mateo, C.M., Maestri, D., Diehl, B.W., Nolasco, S.M., Tomás, M.C. *J Food Comp Anal,* 2011, 24, 166-174.

[15] Capitani, M.I., Spotorno, V., Nolasco, S.M., Tomás, M.C. *LWT - Food Sci Techn*, 2012, 45, 94-102.

[16] Antruejo, A., Azcona, J.O., García, P.T. *Brit Poultry Sci*, 2011, 52 (6), 750-760.

[17] Langman, L., Carduza, F.J., Irurueta, M., Grigioni, G., Gallinger, C. I.. Efecto de la incorporación de semilla de chía (*Salvia hispanica*) en la

dieta de pollos sobre el perfil de aroma de la carne. Congreso Internacional de Ciencia y Tecnología de los Alimentos. Córdoba, Argentina, 2006; pp 345-346.
[18] Ramírez, V.M.L. 2009. Composición de semillas de chía (*Salvia hispanica*) y efecto de la incorporación de su aceite en un producto cárnico. Tesis de Maestría. Universidad Autónoma Metropolitana.
[19] Salazar-Vega, M.I., Rosado-Rubio, J.G., Chel-Guerrero, L.A., Betancur-Ancona, D.A., Castellanos-Ruelas, A.F. *Interc*, 2009, 34(3), 209-213.
[20] Salvador-Vega, L., Gutierrez-Tolentino, R., Coronado-Herrera, M.N., Pérez-González, J.J., Ramírez-Vega, M.L. Adición de aceite de chía (*Salvia hispanica*) como fuentes de ácidos grasos omega 3 en chorizo. In: Avances en la Investigación de la Alimentación Funcional. Editado por. Fontecha-Alonso JF. México, 2010; pp 101-108.
[21] Capitani, M.I., Ixtaina, V.Y., Nolasco, S.M., Tomás, M.C. *J Sci Food Agr*, 2013, 93 (15), 3856-3862.
[22] Lin, K.Y., Daniel, J.D., Whistler, R.L. *Carb Polym,* 1994, 23, 13-18.
[23] Marin Flores, F.M., Acevedo, M.J., Tamez, R.M., Nevero, M.J., Garay, A.L. 2008. WO/2008/0044908 Method for obtaining mucilage from *Salvia hispanica* L. Word Internacional Property Organization.
[24] Hentry, H.S., Mittleman, M., McCrohan, P.R. Introducción de la chía y la goma de tragacanto en los Estados Unidos. En Avances en Cosechas Nuevas. Editado por Janick J, y Simon JE. Prensa de la Madera, Pórtland, O., 1990; pp 252-256.
[25] Hulse, J. Flavor, spices and edible gums: opportunities for integrated agroforesty systems. In: International Conference on Domestication and Commercialization of Non-timber Forest Products in Agroforesty Systems. FAO, Uganda, 1996; p. 298.
[26] Bhatty, R.S. Nutrient composition of whole flaxseed and flaxseed meal. In S. C. Cunnane, & L. U. Thompson (Eds.), Flaxseed in human nutrition Champaign, Illinois: AOCS Press, 1995; pp. 22-42.
[27] Ziolkovska, *Food Hydrocolloid*, 2012, 26, 197-204.
[28] Mazza, G., Biliaderis, C.G. *J Food Sci*, 1989, 54, 1302-1305.
[29] Cui, W., Mazza, G., Oomah, B.D., Biliaderis, C.G. *LWT - Food Sci Techn*, 1994, 27, 363-369.
[30] Wang, Y., Wang, L.J., Li, D., Xue, J., Mao, Z.H. *Carb Polym,* 2009, 78(2), 213–219.
[31] Manson, C.T., Hall, L.A. *Food Ind,* 1948, 20, 382-383.
[32] Stewart, S., Mazza, G. *J Food Quality*, 2000, 23(4), 373–390.
[33] Qin, L., Xu, S.Y., Zhang, W.B. *J Sci Food Agr,* 2005, 85(3), 505–512.

[34] Chen, H.H., Xu, S.Y., Wang, Z. *J Food Eng,* 2007, 80(4), 1051–1059.
[35] Karawya, M.S., Wassel, G.M., Baghdadi, H.H., Ammar, N.M. *Med Phy*, 1980, 38, 73–78.
[36] 36. Dakia, P.A., Bleckerb, C., Roberta, C., Watheleta, B., Paquota, M. *Food Hydrocolloid*, 2008, 22, 807–818.
[37] Bouzouita, N., Khaldi, A., Zgoulli, S., Chebil, L., Chekki, R., Chaabouni, M.M., Thonart, P. *Food Chem*, 2007, 101, 1508-1515.
[38] El Batal, H., Hasib. A., Ouatmane. A., Boulli, A., Dehbi, F., Jaouad, A. *J Mat Env Sci*, 2013, 4(2), 309-314.
[39] Santos, M., Rodrigus, A., Teixeira, J.A. *Bioch Eng J*, 2005, 25(1), 1–6.
[40] Calixto, F.S., Canellas, J. *J Sci Food Agr*, 1982, 33, 1319–1323.
[41] Brummer, Y., Cui, W., Wang, Q. *Food Hydrocolloid*, 2003, 17, 229-236.
[42] Stephen, A.M., Churns, S.C. 1995. Introduction. In A. M. Stephen (Ed.), Food polysaccharides and their application (pp. 1–18). New York: Marcel Dekker.
[43] Garti, N., Madar, Z., Aserin, A., Sternheim, B. *LWT - Food Sci Techn,* 1997, 30, 305-311.
[44] Huang, X., Kakuda, Y., Cui, W. *Food Hydrocolloid,* 2001, 15, 533–542.
[45] Pan. L.G., Tomás. M.C., Añón. M.C. *J Surfact Det*erg, 2002, 5(2), 135-143.
[46] Chau, C., Cheung, K., Wong, Y. *J Agr Food Chem,* 1997, 45, 2500-2503.
[47] 47. Ye, A., Hemar, Y., Singh, H. *Food Hydrocolloid*, 2004, 18, 737-746.
[48] Soleimanpour, M., Koocheki, A., Kadkhodaee, R. *Food Hydrocolloid*, 2013, 30, 292-301.
[49] Wang, Y., Li, D., Wang, L.J., Adhikari, B. *J Food Eng*, 2011, 104, 56-62.
[50] Nor Hayati, I., Bin Che Man, Y., Ping Tan, C., Nor Aini, I. *Food Hydocolloid,* 2009, 23, 233-243.

In: Seed Oil
Editor: Alexis Varnham

ISBN: 978-1-63463-056-6

Chapter 5

Importance of Fatty Acid Composition and Antioxidant Content of Vegetable Oils and Their Blends on Food Quality and Human Health

Estefanía N. Guiotto[1,2], Vanesa Y. Ixtaina[2], Susana M. Nolasco[1] and Mabel C. Tomás[2]

[1]Grupo de Investigaciones TECSE. Departamento de Ingeniería Química. Facultad deIngeniería, UNCPBA, Olavarría, Buenos Aires, Argentina
[2]Centro de Investigación y Desarrollo en Criotecnología de Alimentos (CIDCA), (CCT La Plata –CONICET) Facultad de Ciencias Exactas, Buenos Aires, Argentina

Abstract

The different vegetable oils available on the market for human consumption mainly differ in fatty acid composition. Chia, flaxseed and sacha inchi oils, are sources of fatty acid α-linolenic (ω-3) followed by mustard and canola oils, while sunflower, safflower, corn, soybean and black cumin oils present high linoleic acid content (ω-6). Polyunsaturated fatty acids (PUFA) (ω-3, ω-6) are essential compounds commonly found in vegetable oils. They are nutritionally important for good health and are especially beneficial for individuals suffering from coronary heart disease, diabetes, and immune response disorders. FAO/WHO have recommended that the essential ω-6:ω-3 FA balance in the diet should be between 5:1 and 10:1. This can be achieved by mixing or blending two or more different oils in specific proportions to get a desired fatty acid composition. Blending vegetable oils can increase the levels of bioactive lipids and natural antioxidants in their blends and improve the nutritional value at affordable prices. Oil blends has been a common practice in the many countries. Recently, the manufacture and marketing of blended oils containing common and unconventional edible oils are allowed. This article deals primarily about blends of different vegetable oils in order to obtain products with improved essential ratio in fatty acids (ω-6:ω-3), functional properties and oxidative stability.

INTRODUCTION

Oilseeds are the major source of oils and fatty acids with potential application as nutraceuticals and functional foods. They also might provide low-cost renewable resource of high added value products such as tocopherols and polyphenolic compounds. The oils rich in polyunsaturated fatty acids (PUFA), which are beneficial for human health, and with high level of tocopherols are now added into the infant formulas. Thus, various food products are available as nutraceutical supplements in many countries (Moyad, 2005; Bozan & Temelli, 2008).Vegetable oils are important functional components of foods and have a significant effect on their quality. They do not only contribute to flavor, odor, color, and texture, but also confer a feeling of satiety and palatability of foods. Although vegetable oils constitute only a minor component of foods, their application increases day by day (Rubilar et al., 2012).

The quality of edible oils include organoleptic characteristics -such as flavor, odor, and color for pressed unrefined vegetable oils-, nutritional aspects, oxidative stability, functionality and health features. These characteristics are determined by the chemical composition of the oil. The nutritional attributes of edible oils associated with the presence of minor components and PUFA content play an important role in preventing diseases and improving health. For this reason the formulation of vegetable oil blends with a special composition is important in order to enhance their stability and nutritional value (Frankel & Huang, 1994; Shiela et al., 2004). Blending vegetable oils can increase the levels of bioactive lipids and natural antioxidants in their blends and improve the nutritional value at affordable prices. Recently, the manufacture and marketing of blended oils containing common and unconventional edible oils are allowed.

The relatively short shelf-life of most commercially available vegetable oils limits their use in different applications (Hamed & Abo-Elwafa, 2012). Oxidative rancidity is the primary mechanism affecting stability during storage of processed and packaged vegetable oils (Gulla & Waghray, 2011). Oxidative stability is greatly related with their fatty acid composition and minor components such as tocopherols and tocotrienols. The oxidation process mainly involves the degradation of PUFA and the generation of free radicals, which cause the loss of functional properties and nutritional value (Gordon, 2001; Bozan & Temelli, 2008). Oxidation imparts undesirable flavours and aromas, compromises the nutritional quality of oils, and leads to the induction of toxic compounds (Ramadan & Wahdan, 2012).

Earlier reports on the oxidative stability of individual oils and oil blends indicated that apart from inherent natural antioxidants in oil, PUFA content is an important factor influencing their oxidative stability (Frankel & Huang, 1994; Chu & Kung, 1998). One way to improve the stability of traditional oils (soybean, sunflower and rapeseed) is reducing PUFA content by blending with more saturated or monounsaturated oils, for example *Moringa oleifera* Lam. (Anwar et al., 2007), palm olein (Mobin Siddique et al., 2010), coconut (Bhatnagar et al., 2009), Allam (2001) studied different sunflower oil blends with nine oils on the improvement of the oxidative stability of edible oils. The results revealed a good correlation between the oleic acid content and the oxidative stability of oils.

By other hand, other research works about oil blends were carried out with the main objective of developing nutritional superior oils with recommended fatty acid ratios (ω-6/ω-3 between 5:1 and 10:1) (FAO/WHO, 1997) which are important for human health (Gulla & Waghray, 2011; Mostafa et al., 2013; Chugh & Dhawan, 2014).

VEGETABLE OILS: FATTY ACID COMPOSITION

Most vegetable oils are obtained from fruits, grains or seeds. All oil recovery processes are designed to obtain triglycerides as free as possible from undesirable impurities; to obtain a yield as high as possible consistent with economics of the process; and to produce cake, meal, or flour, usually high in protein content, of maximum value. Three general types of processes are used to crush oilseeds: hard pressing, prepress solvent extraction, and direct solvent extraction. The selection of extraction process depends on the oil content of the source material, the amount of residual oil in the meal allowed, the amount of protein denaturation allowed, the amount of investment capital available, and local environmental laws concerning emissions of volatile organic compounds (Johnson, 2008). Seeds give oils in different proportions; world average oil yields are: soybean (18.3%); rapeseed (38.6%); sunflower (40.9%); groundnut (40.3%); cottonseed (15.1%); coconut (62.4%); palm kernel (44.6%); sesame (42.4%); flaxseed (33.5%) and corn (about 5%) (Gunstone, 2011).

The quality and the potencial use of vegetable oils are mainly determined by their fatty acids composition for two reasons: a) dietary fatty acids profiles have a significant impact on health, and b) fatty acid composition determines the physicochemical characteristics of the oil. According to the number of double-bonded carbons present in their chain, fatty acids are classified into saturated (SFA), monounsaturated (MUFA) and polyunsaturated fatty acids (PUFA). Manipulation of relative concentrations of SFA, MUFA and PUFA allows to obtain different and desired oil properties (Echarte et al., 2010). SFA are more stable than MUFA and PUFA. This stability is important in terms of the shelf life of packaged foods and retardation of rancidity in frying oils. For this purpose, oils rich in MUFA and poor in SFA are preferred because they combine a hypocholesterolemic effect and a high oxidative stability (Echarte et al., 2010). PUFA are essential for humans since they cannot be synthesized in the organism and must be ingested in food. Vegetable oils with the high levels of PUFA, is more readily oxidized if stored or handled improperly. Oxidative stability index is inversely proportional to PUFA content (Frankel & Huang, 1994; Chu & Kung, 1998). The oxidation of PUFA results in the generation of volatile compounds, which are responsible for the off-flavors in the food industry. In addition, α-linolenic acid belongs to the ω-3 family which is an essential component of healthy food (Rubilar et al., 2012). The International Food and Nutrition Committees convened by FAO/WHO (1997) have established that fats in general should not contribute more than 30% of total calories consumed by an adult. Furthermore, they recommend that the distribution of consumption of different types of fatty acids may be within 30%, a contribution of 10% from SFA, 10% of the MUFA and about 10% of PUFA. This is a 1:1:1 relationship between SFA, MUFA and PUFA. Also, it has been suggested that the ω-6/ω-3 fatty acids ratio in the diet should be between 5:1 and 10:1. Individuals who consume a ratio in excess of 10:1 should be encouraged to eat more ω-3 rich foods (FAO/WHO, 1997).

Table 1. Fatty acids composition in some vegetable oils

Oil	Fatty acids (%)						Reference
	$C_{16:0}$	$C_{18:0}$	$C_{18:1}$	$C_{18:2\ (\omega\text{-}6)}$	$C_{18:3\ (\omega\text{-}3)}$	ω-6/ω-3 ratio	
Black cumin	13	2.9	21.1	57.7	nd	nd	Hamed & Abo-Elwafa (2012)
Canola	5.5	3.4	54.8	24.9	9.9	2.5	Mostafa et al. (2013)
Chia	7.1	2.1	6.3	19.4	65.2	0.3	Guiotto et al. (2014)
Coconut	8.4	2.8	6.1	1.2	nd	nd	Bhatnagar et al. (2009)
Corn	10.7	1.6	24.5	61.3	1.1	55.7	Ramadan (2013)
Flaxseed	4.1	4.6	28.2	20.4	41.0	0.5	Mostafa et al. (2013)
Groundnut	15.9	1.4	46.2	36.1	nd	nd	Sunil et al. (2013)
HOSUN	4.7	3.6	64.5	24.8	nd	nd	Chu & Kung (1998)
Mustard	2.7	nd	10.5	14.3	11.4	1.2	Chugh & Dhawan (2014)
Palm	43.3	4.8	42.4	7.8	nd	nd	Bhatnagar et al. (2009)
Rice bran	19.1	3.9	40.5	35.6	nd	nd	Mishra et al. (2012)
Sacha Inchi	4.3	3.0	9.0	36.2	46.8	0.8	Fanali et al. (2011)
Safflower	5.8	1.0	20.0	69.9	nd	nd	Mishra et al. (2012)
Sesame	12.7	0.9	40.3	45.8	nd	nd	Sunil et al. (2013)
Soybean	10.4	3.5	21.5	51.5	7.8	6.6	Ramadan (2013)
SUN	6.6	2.3	36.6	54.4	nd	nd	Guiotto et al. (2014)

SUN: Sunflower oil; HOSUN: High Oleic Sunflower oil; nd: nodetected.

The ω-3 fatty acids are significant structural components of the phospholipid membranes of tissues throughout the body and are especially rich in the retina, brain, and spermatozoa. Another important feature of ω-3 fatty acids is their roles in the modulation and prevention of human diseases, particularly coronary heart disease. The antiarrhythmic effect of ω-3 fatty acids is a discovery that has great relevance to the prevention of sudden death from ventricular fibrillation. Certainly, the evidence is now strong that ω-3 fatty acids are essential for human development in uterus and in infancy and are likely to have a role throughout life (Connor, 2000).

Fatty acid composition of different vegetables oils are presented in Table 1. It can be seen that no single conventional oil presented the desired ideal fatty acid ratio recommended by health agencies i.e. 1:1:1 SFA:MUFA:PUFA. Sunflower *(Helianthus annuus* L.*)* oil, which is a non-genetically modified (non-GMO) source of vegetable oil, is available with three different fatty acid compositions: traditional sunflower varieties, with linoleic acid (ω-6) content of 65-70%; high-oleic (HOSUN) varieties, with > 80% of oleic acid and 5–9% of linoleic acid; and mid-oleic varieties, with 55–75% of oleic acid and 15–35% of linoleic acid (Gunstone, 2011). Sunflower seed oil has a very low content of α-linolenic acid. This fact gives some increased oxidative stability but does not provide valuable ω-3 fatty acids needed for health nutrition (List, 2014). There are other vegetable oils with high linoleic acid content (ω-6) such as safflower, corn, soybean and black cumin oils. Regarding ω-3, chia (*Salvia hispanica* L.*)*, flaxseed (*Linum usitatissitmum* L.) and Sacha Inchi (*Plukenetia volubilis* L.) oils contain the highest proportion of α-linolenic acid of any known vegetable sources, followed by mustard (*Sinapis alba* L.), soybean (*Glycine max* L.) and canola (*Brassica napus* L.) oils (Table 1).

Palm oil has a balanced fatty acid composition in which the level of saturated fatty acids is almost equal to that of the unsaturated fatty acids. Palmitic and oleic acids are the major component of this oil, followed by linoleic acid and stearic acids and only traces of α-linolenic acid. The low level of PUFA makes this oil relatively stable to oxidative deterioration. Coconut oil is rich in SFA (~93%), mainly medium chain fatty acids ($C_{6:0}$, $C_{8:0}$, $C_{10:0}$, $C_{12:0}$) (~60%), and especially $C_{12:0}$ (~50%) (Bhatnagar et al., 2009). Mustard oil contains more than 50% of erucic acid ($C_{22:1}$) which is higher than the desirable and internationally accepted level of < 5% (Chugh & Dhawan, 2014). Rice bran oil comprises about 20% of SFA and an approximately constant balance of MUFA and PUFA. Another interesting feature of rice bran oil is its high unsaponifiable matter content compared to other oils (Mezouari

& Eichner, 2007). Groundnut oil has a fatty acid composition similar to that of rice bran oil (Gunstone, 2011).

Soybean oil is one of the major cooking oils. However, the high linolenic acid content causes oil instability at high temperatures (Chu & Kung, 1998).

The fatty acid composition of canola oil presents low levels of SFA (8-9%), high levels of the MUFA (oleic acid, 55%) and moderate PUFA content (Eskin & McDonald, 1991; Mostafa et al. (2013). According to the *Codex Alimentarius* (2013), low-erucic acid rapeseed oil must not contain more than 2% erucic acid.

Sesame oil is classified as polyunsaturated, semi-drying oil containing about 86% of unsaturated fatty acids. The fatty acid composition in sesame oil is mainly characterized by equal proportion of oleic acid and linoleic acid, small amounts of saturated acids, and only a little α-linolenic acid content (Gunstone, 2011).

The possibility of developing nutritionally more suitable oils with recommended fatty acid ratios can be carried out using different edible oils and blending them to improve the fatty acid balance. Table 2 shows the fatty acid composition of some oil blends reported in the literature. Fatty acids differ according to the type and proportion of oil used in the formulation. According to Guiotto et al. (2014), the fatty acid composition corresponding to sunflower-chia oil blends indicates that the essential fatty acids balance ω-6:ω-3 (5:1 to 10:1) can be achieved with a low proportion of chia oil (10 and 20% wt/wt). Mostafa et al. (2013) also prepared flaxseed and canola oil blends according to FAO/WHO recommendation.

Refined vs. Cold Pressed Oils and Natural Antioxidant Content

Some oils are used without further treatment but most are refined before use. The refining processes remove undesirable materials (phospholipids, mono and diacylglycerols, free fatty acids, pigments, oxidized materials, flavour components, trace metals and sulfur compounds) but may also eliminate valuable minor components which are antioxidants and vitamins such as carotenes and tocopherols (Gunstone, 2011). Thus, it can be very interesting to carry out partial refinements which kept or leave a remnant of carotenoids, tocopherols and phytosterols, which will add more nutritional value to the product.

Table 2. Fatty acids composition of some oil blends

Oil blends	Ratio (wt/wt)	Fatty acids (%)						Reference
		$C_{16:0}$	$C_{18:0}$	$C_{18:1}$	$C_{18:2}$ (ω-6)	$C_{18:3}$ (ω-3)	ω-6/ω-3 ratio	
Canola:Flaxseed	19:1	5.4	3.5	52.1	24.8	13.0	1.9	Mostafa et al. (2013)
Corn:Black cumin	8:2	10.1	2.9	25.6	59.5	0.3	198.3	Ramadan & Wahdan (2012)
Flaseed:Black cumin	8:2	5.7	3.1	19.1	16.6	53.3	0.31	Hamed & Abo-Elwafa (2012)
Flaxseed:Canola	1.3:1	4.7	4.0	40.1	22.4	27.0	0.8	Mostafa et al. (2013)
Flaxseed:Sesame	8:2	5.9	5.6	18.6	16.9	52.9	0.32	Hamed & Abo-Elwafa (2012)
Groundnut:Rice bran	8:2	19.0	0.5	45.5	34.8	nd	nd	Sunil et al. (2013)
Groundnut:Sesame	8:2	16.0	0.7	45.4	37.8	nd	nd	Sunil et al. (2013)
Mustard:Sesame	8:2	5.7	5.4	20.4	24.8	0.6	41.3	Gulla & Waghray (2011)
Rice Bran:Sesame	8:2	13.3	2.8	38.3	39.2	0.7	55.7	Gulla & Waghray (2011)
Safflower:Rice bran	8:2	7.5	2.0	23.1	64.1	0.1	641.0	Mishra et al. (2012)
Sesame:Mustard	8:2	5.5	5.5	20.2	24.9	0.6	43.2	Gulla & Waghray (2011)
Sesame:Rice bran	8:2	16.5	3.9	45.6	29.7	nd	nd	Gulla & Waghray (2011)
Soybean:HOSUN:SUN	8:1:1	9.4	3.5	25.7	50.5	6.3	8.0	Chu & Kung (1998)
Soybean:HOSUN:SUN	1:8:1	5.4	3.6	55.8	31.8	1.0	31.8	Chu & Kung (1998)
Soybean:HOSUN:SUN	1:1:8	6.5	3.5	25.2	62.0	1.0	62.0	Chu & Kung (1998)
SUN:Black cumin	8:2	7.9	3.8	26.1	60.0	0.3	200.0	Ramadan (2013)
SUN:Chia	8:2	7.6	2.3	26.1	46.7	17.4	2.7	Guiotto et al. (2013)
SUN:Chia	9:1	7.3	1.1	34.6	48.0	9.0	5.3	Guiotto et al. (2013)
SUN:Flaxseed	6.5:3.5	6.6	2.7	23.6	47.5	19.2	2.4	Umesha & Naidu (2012)
SUN:Rice bran	8:2	10.4	2.3	42.0	50.6	0.2	253.5	Mishra et al. (2012)

SUN: Sunflower oil; HOSUN: High Oleic Sunflower oil; nd: no detected.

Over the last few years, there is an increased interest in cold pressed oils due to their high nutritive value (bioactive lipids, natural antioxidants). The cold pressing procedure is becoming an interesting substitute for conventional practices because of consumers' desire for natural and safe food products (Lutterodt et al., 2010). Cold pressing is a technology for seed oil production, which involves no heat oils or chemical treatments. Cold pressing did not involve any refining process and the obtained oil may contain a high level of lipophilic phytochemicals such as natural antioxidants. Thus, crude vegetable oils are usually oxidatively more stable than the corresponding refined and processed oils. The oxidative stability depends on the fatty acid composition and the presence of minor components such as tocols, carotenoids, metal ions, polar lipids and the initial amount of hydroperoxides (Ramadan, 2013).

Oilseeds present natural substances with antioxidant properties such as tocopherols, β-carotene, oryzanol and lignans. In recent years, with the growing health awareness among consumers, the antioxidants of vegetable oils are being isolated and used as nutritional supplements. These minor constituents are associated with medicinal aspects, such as preventing/delaying onset of diseases and promoting health (Sunil et al., 2013). Palm oil is an important source of β-carotene, which functions as provitamin-A and a scavenger of oxygen free radicals (Basu et al. 2001). Oryzanol, only present in rice bran oil (RBO), is a mixture of at least five sterol esters of ferulic acid, which is shown to have hypocholesterolemic activity (Mezouari & Eichner, 2007; Reena & Lokesh, 2007). Furthermore, RBO contains other high-value compounds including tocotrienols and squalene. Sesame oil is a rich source of lignans (sesamin and sesamolin), which are known to have antioxidant, hepato protective, hypolipidemic, hypotensive and anticarcinogenic activities (Namiki, 2007).

Tocopherols (present in all vegetable oils) exist in four different naturally-occurring forms (α-, β-, γ- and δ-tocopherol) that differ in the location of the methyl groups on the chromanol ring. There are differences among the four types of tocopherols in relation with their antioxidant activity *in vitro* and *in vivo*. Thus, α-tocopherol is characterized by a maximum effectiveness as in vivo antioxidant or vitamin E, but its *in vitro* activity is low in comparison with other tocopherols. In contrast, γ-tocopherol has a high in vitro antioxidant activity. Tocopherols act as antioxidant by donating an hydrogen atom to chain propagating peroxil radicals. Tocopherols in vegetable oils are believed to protect PUFA from peroxidation (Ramadan & Wahdan, 2012). Moreover, tocopherols are the major lipid-soluble, membrane-localized antioxidants in humans. Epidemiologic studies suggest that vitamin E deficiency may result from malnutrition and bad intestinal absorption syndromes or other pathologies associated with poor absorption of fats. It produces membrane fragility in human red blood cells, while long term deficiency is thought to result in neurological dysfunction. Tocopherol requirements in humans are believed to depend on dietary content of PUFA; a requirement of 0.6 mg tocopherol per g PUFA has been suggested (Pereyra-Irujo & Aguirrezábal, 2007).

Table 3. Tocopherols content of different vegetable oils

Oil	Tocopherols (mg/kg)					Reference
	α-	β-	γ-	δ-	Total	
Black cumin	64.2	53.9	208.9	14.5	341.5	Ramadan (2013)
Canola	170	134	403	41	748	Chu & Kung (1998)
Chia	nd	nd	404	7	411	Guiotto et al. (2014)
Corn	417	19.7	419	22	879	Ramadan & Wahdan (2012)
Flaxseed	12	nd	520	9.5	541.5	Schwartz et al. (2008)
HOSUN	787	nd	66	nd	853	Chu & Kung (1998)
Rice bran	65	109		7	181	Umesha & Naidu (2012)
Sesame	79	4.1	360	12	455.1	Schwartz et al. (2008)
Soybean	144	27	624	229	1025	Chu & Kung (1998)
Sunflower	498	4	nd	nd	502	Guiotto et al. (2014)

nd: no detected.

Table 4. Tocopherol content of vegetable oil blends

Oil blend	Ratio wt:wt	Tocopherols (mg/kg)				Total	Reference
		α-	β-	γ-	δ-		
SUN:Flaxseed	6.5:3.5	241	160		9	410	Umesha & Naidu (2012)
Corn:Black cumin	9:1	402	20.8	401	25.4	850	Ramadan & Wahdan (2012)
Corn:Black cumin	8:2	387	21.8	383	28.6	821	Ramadan & Wahdan (2012)
SUN:Chia	8:1	376	6	72	nd	454	Guiotto et al. (2014)
SUN:Chia	9:1	422	2	9	nd	433	Guiotto et al. (2014)

nd: no detected.

Results of qualitative and quantitative composition of tocopherols in vegetable oils and their blends are summarized in Tables 3 and 4, respectively. It is possible to observe that α and γ-tocopherols present the higher level in vegetables oils. The major tocopherol present in chia, flaxseed, canola, soybean and sesame is γ-tocopherol, while sunflower oil contained a high amount of α-tocopherol, and corn oil shown a balance between α and γ-tocopherols. The profile of tocopherols in oil blends can be enriched through the mixture with oils obtained by cold pressing (Ramadan, 2013).

CONCLUSION

Different studies about the importance of fatty acid composition and antioxidant content of vegetable oils and their blends on food quality and human health have been carried out. It is well recognized that the polyunsaturated fatty acids content is an important factor influencing oil stability and quality. Vegetable oil blends are being developed to improved their nutritional profile. PUFA rich oils (mainly ω-3), such as chia, flaxseed and sacha inchi oils, can be blended to enrich vegetables oils to obtain a balanced ω-6/ω-3 ratio according to FAO/WHO recommendation. Moreover, the inclusion of oils partially refined in oil blends can be very interesting due to the contribution in a certain level of carotenoids, tocopherols and phytosterols, resulting in more nutritional added value products.

REFERENCES

Allam, S. H. (2001). Utilization of some untraditional sources of high oleic acid oils for improving vegetable oils stability. *Rivista Italiana Delle Sostanze Grasse*, *78* (6), 337–341.

Anwar, F., Hussain, A. I., Iqbal, S. & Bhanger, M. I. (2007). Enhancement of the oxidative stability of some vegetable oils by blending with *Moringa oleifera* oil. *Food Chemistry*, *103* (4), 1181-1191.

Basu, H. N., Vecchio, A. J. D., Flider, F. & Orthoefer, F. T. (2001). Nutritional and potential disease prevention properties of carotenoids. *Journal of the American Oil Chemists' Society*, *78*, 665–675

Bhatnagar, A. S., Kumar, P. P., Hemavathy, J. & Krishna, A. G. (2009). Fatty acid composition, oxidative stability, and radical scavenging activity of vegetable oil blends with coconut oil. *Journal of the American Oil Chemists' Society*, *86*(10), 991-999.

Bozan, B. & Temelli, F. (2008). Chemical composition and oxidative stability of flax, safflower and poppy seed and seed oils. *Bioresource Technology*, *99* (14), 6354-6359.

Chu, Y. H. & Kung, Y. L. (1998). A study on vegetable oil blends. *Food Chemistry*, *62*(2), 191-195.

Chugh B. & Dhawan K. (2014). Storage studies on mustard oil blends. *Journal of Food Science and Technology*, *51* (4), 762-767.

Codex Alimentarius Commission (2013). Codex Stan 210-1999. Codex Standard for named vegetable oils. www.codexalimentarius.org/input/download/.../CXS_210e.pdf (accessed August 2014)

Connor W. E. (2000). Importance of n−3 fatty acids in health and disease. *The American Journal of Clinical Nutrition*, *71*(1), 171S-175S.

Echarte, M. M., Pereyra-Irujo, G. A., Covi, M., Izquierdo, N. G. & Aguirrezábal, L. A. N. (2010). Producing better sunflower oils in a changing enviroment. in: *Advances in Fats and Oils Research*. Ed, Tomás Mabel. Chapter 1, 1-23.

Eskin, N. M. & McDonald, B. E. (1991). Canola oil. *Nutrition Bulletin*, *16*(3), 138-146.

Fanali, C., Dugo, L., Cacciola, F., Beccaria, M., Grasso, S., Dachà, M. & Mondello, L. (2011). Chemical characterization of Sacha Inchi (*Plukenetia volubilis* L.) oil. *Journal of Agricultural and Food Chemistry*, *59*(24), 13043-13049.

FAO/WHO Organización Mundial de la Salud. (1997). Grasas y Aceite en la Nutrición Humana. Consulta FAO/OMS de expertos. Roma. http://www.fao.org/docrep/V4700S/v4700s00.htm (accessed August 2014)

Frankel, E. N. & Huang, S. W. (1994). Improving the oxidative stability of polyunsaturated vegetable oils by blending with high-oleic sunflower oil. *Journal of the American Oil Chemists' Society*, *71*(3), 255-259.

Gordon, M. (2001). The development of oxidative rancidity in foods. In: Pokorný, J., Yanishlieva, N., Gordon, M. (Eds.), *Antioxidants in Food*. CRC Press, Boca Raton, 7–21.

Guiotto, E. N., Ixtaina, V. Y., Nolasco, S. M. & Tomás, M. C. (2014). Effect of Storage Conditions and Antioxidants on the Oxidative Stability of Sunflower–Chia Oil Blends. *Journal of the American Oil Chemists' Society*, *91* (5), 767-776.

Gulla, S. & Waghray, K. (2011). Effect of storage on physico-chemical characteristics and fatty acid composition of selected oil blends. *European Journal of Lipid Science and Technology*, *3*, 35-46.

Gunstone, F. (Ed.) (2011). *Vegetable oils in food technology: composition, properties and uses*. John Wiley & Sons.

Hamed, S. F. & Abo-Elwafa, G. A. (2012). Enhancement of oxidation stability of flax seed oil by blending with stable vegetable oils. *Journal of Applied Sciences Research*, *8*(10), 5039-5048.

Johnson, A. (2008). Recovery, Refining, Converting, and Stabilizing Edible Fats and Oils. In: *Food Lipids. Chemistry, Nutrition and Biotechnology*. Ed. Akoh CC & Min DB. CRC press, Boca Raton, FL.

List, G. (2014). Sunflower seed and oil. *Lipid Technology*, *26* (1), 24-24.

Lutterodt, H., Luther, M., Slavin, M., Yin, J.-J., Parry, J., Gao, J.-M., Yu, L. (2010). Fatty acid profile, thymoquinone content, oxidative stability, and antioxidant properties of cold pressed black cumin seed oils. *LWT- Food Science and Technology*, *43*, 1409–1413.

Mezouari, S. & Eichner, K. (2007). Evaluation of the stability of blends of sunflower and rice bran oil. *European Journal of Lipid Science and Technology*, *109*(5), 531-535.

Mishra, R., Sharma, H. K. & Sengar G. (2012). Quantification of rice bran oil in oil blends. *Grasas y Aceites*, *63*(1), 53-60.

Mobin Siddique, B., Ahmad, A., Hakimi Ibrahim, M., Hena, S., Rafatullah, M., Omar, A. K. M. (2010). Physico-chemical properties of blends of palm olein with other vegetable oils. *Grasas y Aceites*, *61*(4), 423-429.

Mostafa, R. A., Moharram, Y. G., Attia, R. S., Sharnouby, S. E. (2013). Formulation and characterization of vegetable oil blends rich in omega–3 fatty acids. *Emirates Journal of Food and Agriculture*, *25* (6), 426-433.

Moyad M. A. (2005). An introduction to dietary/supplemental ω-3 fatty acids for general health and prevention: Part I. *Urologic Oncology: Seminars and Original Investigations*, *23* (1), 23–35.

Namiki M. (2007). Nutraceutical functions of sesame: A review. *Critical Reviews in Food Science and Nutrition*, *47*, 651–673.

Padmavathy A., Siddhu A. & Sundararaj P. (2001). Effect of blending edible grade crude palm oil with refined groundnut or sunflower oils on storage stability and sensory attributes. *Journal of the Oil Technologists' Association of India*, *33*(3), 93–103.

Pereyra-Irujo, G. A. & Aguirrezábal, L. A. (2007). Sunflower yield and oil quality interactions and variability: Analysis through a simple simulation model. *Agricultural and Forest Meteorology*, *143* (3), 252-265.

Ramadan, M. F. (2013). Healthy blends of high linoleic sunflower oil with selected cold pressed oils: Functionality, stability and antioxidative characteristics. *Industrial Crops and Products*, *43*, 65-72.

Ramadan M. F. & Wahdan, K. M. M. (2012). Blending of corn oil with black cumin (*Nigella sativa*) and coriander (*Coriandrum sativum*) seed oils: Impact on functionality, stability and radical scavenging activity. *Food Chemistry*, *132* (2), 873-879.

Reena B. M. & Lokesh B. R. (2007). Hypolipidemic effect of oils with balanced amounts of fatty acids obtained by blending and interesterification of coconut oil with rice bran oil or sesame oil. *Journal of Agricultural and Food Chemistry*, *55*, 10461–10469.

Rubilar M., Morales E., Sáez R., Acevedo F., Palma B., Villarroel M. & Shene C. (2012). Polyphenolic fractions improve the oxidative stability of microencapsulated linseed oil. *European Journal of Lipid Science and Technology*, *114*(7), 760-771.

Schwartz H., Ollilainen V., Piironen V. & Lampi, A. M. (2008). Tocopherol, tocotrienol and plant sterol contents of vegetable oils and industrial fats. *Journal of Food Composition and Analysis*, *21*(2), 152-161.

Shiela, P. M., Sreerama, Y. N. & Krishna, A. G. (2004). Storage stability evaluation of some packed vegetable oil blends. *Journal of the American Oil Chemists' Society*, *81* (12), 1125-1129.

Sunil L., Reddy, P. V., Krishna, A. G. & Urooj, A. (2013). Retention of natural antioxidants of blends of groundnut and sunflower oils with minor oils during storage and frying. *Journal of Food Science and Technology*, 1-9.

Umesha S. S. & Naidu A. K. (2012). Vegetable oil blends with a-linolenic acid rich Garden cress oil modulate lipid metabolism in experimental rats. *Food Chemistry*, *135*, 2845–2851.

In: Seed Oil
Editor: Alexis Varnham

ISBN: 978-1-63463-056-6

Chapter 6

ELIMINATION OF TOXIC PHORBOL ESTERS IN *JATROPHA CURCAS* SEED OIL BY ADSORPTION TECHNIQUE

Vittaya Punsuvon[1,2,*] and Rayakorn Nokkaew[1]

[1]Department of chemistry, Faculty of Science,
Kasetsart University, Bangkok, Thailand
[2]Center of Excellence-Oil Palm,
Kasetsart University, Bangkok, Thailand

ABSTRACT

Nowadays *Jatropha curcas* is one of the important alternative oil plants to produce biodiesel. But because of toxic substance especially phorbol esters are dangerous compounds for human who working with this oil. And so it need to eliminate this substance before utilization.

Phorbol esters are a natural toxic ester found in tropical plant in the family of Euphorbiaceae. It is main toxic compounds in seed oil of Jatropha curcas. The biological effects of phorbol esters are tumor promotion or cocarcinogen when taken and inflammation when contacted. At least 5 types of phorbol esters are detected in *J. curcas* oil. The major chemical structure of detected phorbol ester is 12-Deoxy-16-hydroxyphorbol-4'-[12',14'-butadienyl]-6'-[16',18'20-nonatrie-nyl]-

---

*Corresponding author: Vittaya Punsuvon. Department of chemistry, Faculty of Science, Kasetsart University, Bangkok 10900, Thailand. Center of Excellence-Oil Palm, Kasetsart University, Bangkok 10900, Thailand. E-mail: fscivit@ku.ac.th.

bicyclo[3.1.6]hexane-(13-0)-2'-[carboxylate]-(16-0)-3'-[8'-butenoic-10'] ate or DHPB.

Many researchers tried to detoxify phorbol esters in seed oil by the extraction with ethanol or methanol but this experiment is difficult to apply for industrial scale because of the immense solvent consumption. Some researcher studied on tradition oil refining process by using deacidification followed bleaching step. The result of experiment showed only 55% of phorbol esters were removed. So in our experiment, the adsorption technique using bentonite was applied to adsorp phorbol esters compounds. The result showed that the optimum adsorption condition on *J. curcas* oil was 3.2%(w/v) of bentonite, 15 min of adsorption time, 100 rpm of stirring rate at room temperature. The phorbol esters can be removed up to 98% for one time of adsorption. This technique is recommended for detoxification *J. curcas* oil in large scale production.

In addition, our study also develop a technique to confirm the presence of phobol esters left in oil after adsorption using liquid chromatography-tandem mass spectrometry with multiple reaction monitoring mode that detects the ionization of parent molecule with mass 711 to precursor and product ion with mass 311 and 293 respectively. This technique is useful technique to confirm phorbol esters left in oil.

Keywords: Phorbol esters, Adsorption, Biodiesel, *Jatropha curcas*

1. INTRODUCTION

Nowadays, the demand and supply gap of vegetable oil has been widening all over the world because of the oil price is increased. Globally, the usage of friendly environmentally fuels is encouraged.

The energy extracted from biomass and tree based materials are perhaps the oldest source of renewable energy. Biomass can be generated from various sources, such as edible and non-edible seed oils, algae and bacteria, forest residues, waste from food and processing, kitchen wastes, etc. The most important biofuels generated from biomass are biodiesel and bioethanol.

Thailand is not rich in petroleum reserves and crude oil, petroleum products must be imported to meet growing energy needs. These fuel and products are usually high prices.

The seeking alternative energy is urgently needed for biodiesel production. Plant species which can be processed to provide a diesel fuel substitute have captured the interest of Thai scientists.

Most of these plant species are such as palm, coconut, soy bean, sunflower, *Jatropha curcas* L. (Saboodum), etc. Ministry of Thai Energy has a policy on renewable energy strategy in the year 2004 that the use of renewable energy in Thailand will increase about 8% of the total energy or 6,540,000 tons within the year 2011 which biodiesel is the one purpose of renewable energy.

Thai government has a policy to support *J. curcas* L. plantation for farmers mainly for renewable energy. *J. curcas* L. is a drought-resistant shrub. It is a member of Euphobiaceae family which is cultivated in Central and South America, South-east Asia, India and Africa. This plant came to Thailand about 200 years ago by Portuguese. Seed oil of *J. curcas* L. is used for soap making and lighting for lamps. The plants grow quickly, survive in poor stony soil and resist to drought. The height of the plant is 2-7 meters and the lifetime is about 50 years. In Thailand the name Saboodam is usually used for *J. curcas* L. The plant can be used in many ways, such as to prevent erosion, reclaim land, grown as a live fence, etc. The seed kernels contain 40-60% oil (Makkar et al., 1997) in which its fatty acid composition is similar to the oil used for human nutrition (Gübitz et al. 1998). A total of 19-27% crude protein can be obtained from press cake (Makkar et al., 1997) which can be a protein source for animal feed. The kernels also contain a number of several toxic and antinutritional compounds. These compounds are trypsin inhibitors, lectins, saponins, phytate and phorbol esters which might cause or at least aggravate the adverse effects in the long term contact, except phorbol esters affect on the short term contact (Makkar et al., 1997).

Phorbol esters are toxic substances that found in plant species of Euphobiaceae and Thymelaceae families. Their structures are based on tetracyclic carbon skeleton known as tigliane. They are known to cause a wide range of biological effects including tumor promotion, cell proliferation, activation of blood plateles and inflammation (Aitken, 1986). These effects are closely related to the structure of several compounds.

Therefore, detoxification of these phorbol esters from the seed oil is required, even when it is industrially used because of the possibility to direct contact of persons with the seed oil. Many experiments eliminate phorbol esters in seed oil by the extraction with ethanol (Gross et al., 1997).

This experiment is difficult to apply for industrial scale because of the immense solvent consumption. Experiment on traditional oil refining process that examines the effects on the phorbol esters content from *J. curcas* oil was performed by Hass (Hass, 2000). It showed that deacidification step and bleaching step could reduce the content of phorbol esters up to 55%.

In addition, phorbol esters are heat stable and can withstand roasting temperature as high as 160°C for 30 min (Makkar and Becker, 1997).

In this experiment, the adsorption process of phorbol esters from *J.* curcas seed oil is examined. In seed oil, bleaching steps in refining of edible oil process can be replaced by the adsorption process.

2. LITERATURE REVIEW

2.1. *Jatropha Curcas* Linn

2.1.1. Botanical Description

Jatropha curcas L., as known as 'physic nut, purging nut, big purging nut, American purging nut, black vomit nut, saboodum, etc.', is a member of the Euphobiaceae family.

It is a tropical plant which can reach a height of 2-7 meters. It is cultivated mainly as a hedge in many Latin America, Asia and African countries. It can be grown in low and high rainfall areas either in the farms as a commercial crop or on the boundaries as a hedge to protect fields from grazing animals and to prevent erosion.

2.1.2. Utilization of Various Parts of Jatropha curcas L.

All parts of *J. curcas* L. have been used in traditional medicine and for various purposes. The oil has been used as a purgative, to treat skin diseases and to soothe pain such as rheumatism. Decoction of the leaves has been used against coughs or as antiseptics after birth, and the branches as chewing sticks (Heller, 1996).

Various extracts from *Jatropha* seeds and leaves show molluscicidal, insecticidal and fungicidal properties (Nwosu and Okafor, 1995; Liu et al., 1997; Solsoloy et al., 1997). The utilization of various parts of *J. curcas* L. is reviewed in Figure 1 (Gübitz et al., 1999).

2.1.3. Chemical and Physical Properties of Jatropha curcas L.

The seed kernels, which seem to be the part of the plant with the highest potential for utilization, contain 40-60% oil (Makkar et al., 1997) with a fatty acid composition similar to oils used for human nutrition (Gübitz et al., 1999).

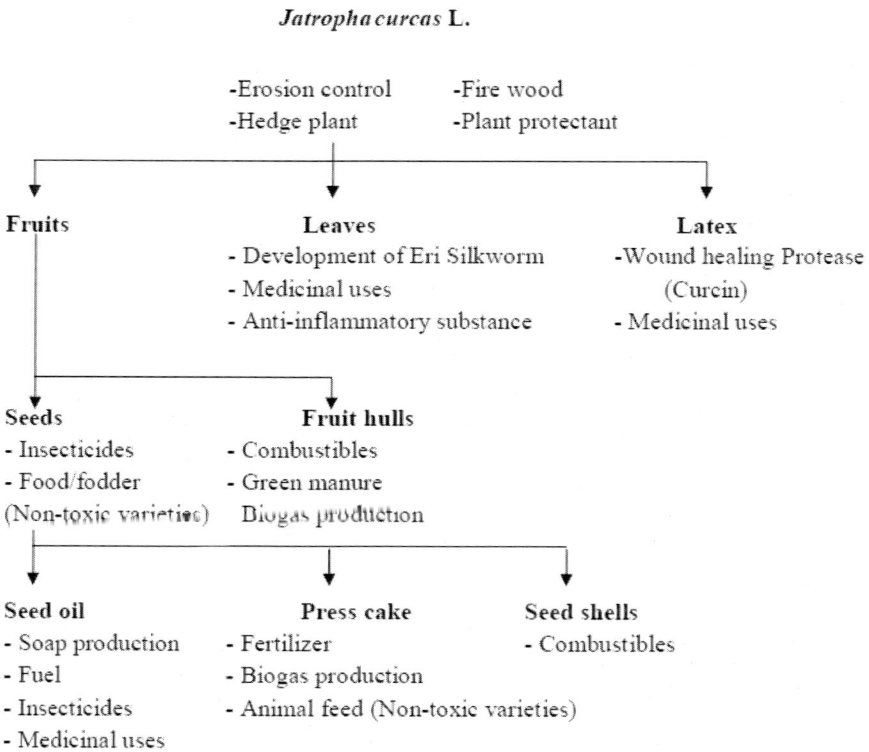

Source: Gübitz et al. (1999).

Figure 1.Exploitation of *Jatropha curcas* L.

2.2. Phorbol Esters

Phorbol esters have been identified as the major toxic principal in *J. curcas* L. (Makkar and Becker, 1997). Phorbol esters were first isolated in 1934 as the hydrolysis product of *Croton tiglium* oil and its structure was determined in 1967. Later, phorbol esters analogues are found in several members of the plant family Euphorbiaceae and *J. curcas* L. is also the plant in family Euphorbiaceae. Phorbol esters in *Jatropha* kernels content at least four different types which can cause the short term toxicity (Makkar et al., 1999). The main chemical structure of phorbol esters in *Jatropha* kernel is 12-Deoxy-16-hydroxyphorbol-4'-[12',14'-butadienyl]-6'-[16',18',20'-nonatrienyl]-bicyclo[3.1.0]hexane-(13-*0*)-2'-[carboxylate]-(16-*0*)-3'-[8'-butenoic-10']ate (DHPB) as shown in Figure 2.

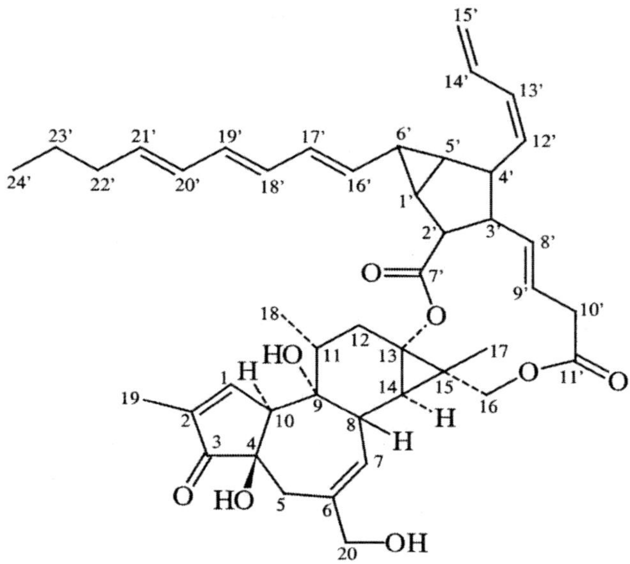

Source: Hass and Mittelbach (2000).

Figure 2. Structure of 12-Deoxy-16-hydroxyphorbol-4'-[12', 14'-butadienyl]-6'-[16', 18', 20'-nonatrienyl]- bicyclo [3.1.0]hexane-(13-*0*)-2'-[carboxylate]-(16-*0*)-3'-[8'-butenoic-10']ate; (DHPB).

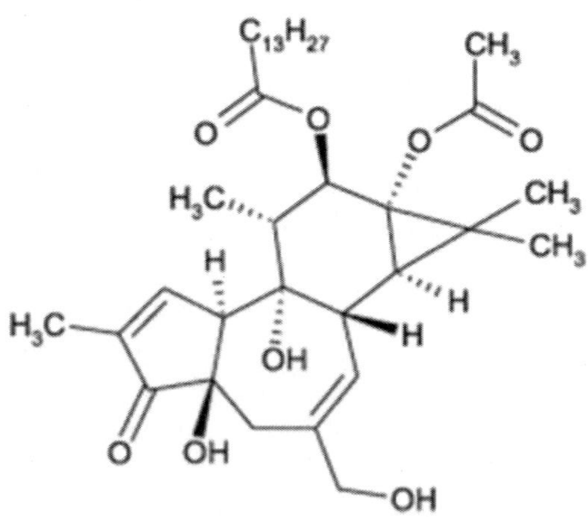

Figure 3.Structure of phorbol-12-myristate 13-acetate or 12-*O*-tradecanoylphorbol13-acetate; (TPA).

Figure 3 shows the chemical structure of phorbol-12-myristate 13-acetate (TPA). It is the phorbol esters standard found in the commercial market and used as phorbol esters standard in this experiment. Mitsuru et al. (1988) reports the tumors-promoting activity of DHPB. It is weaker than TPA because the application of 2.5 μg of TPA induces tumors nearly 100% in mice within 12 weeks. DHPB results in 46.7% incidence of tumors within 30 weeks. The weaker activity of DHPB might be explained by the structural difference between DHPB and TPA. They contain (a) the alcohol moiety that is 12-deoxy-16-hydroxy phorbol of DHPB and TPA, (b) the acid moieties that is the unsaturated acid of DHPB and saturatured acid of TPA.

2.2.1. Definition of Phorbol Esters

The fundamental substance of phorbol esters is the alcohol moiety, of this family of compounds is tigliane, a tetracyclic diterpene. Hydroxylation of this fundamental substance in various positions and connection to various acid moieties by ester bonding characterize the large number of compounds termed as phorbol esters (Evans, 1986), as shown in Figure 4.

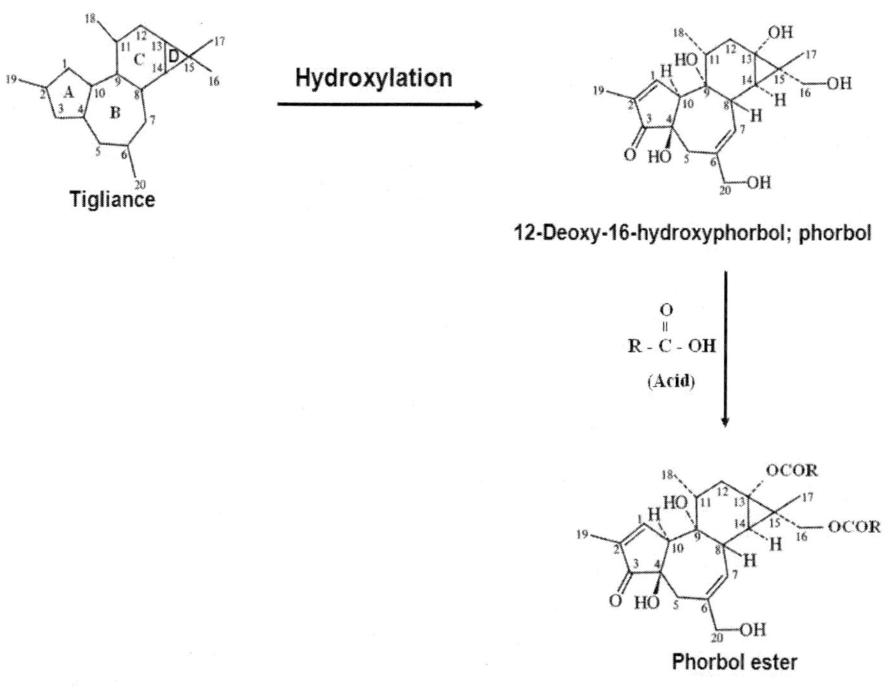

Figure 4.Occurring of phorbol esters.

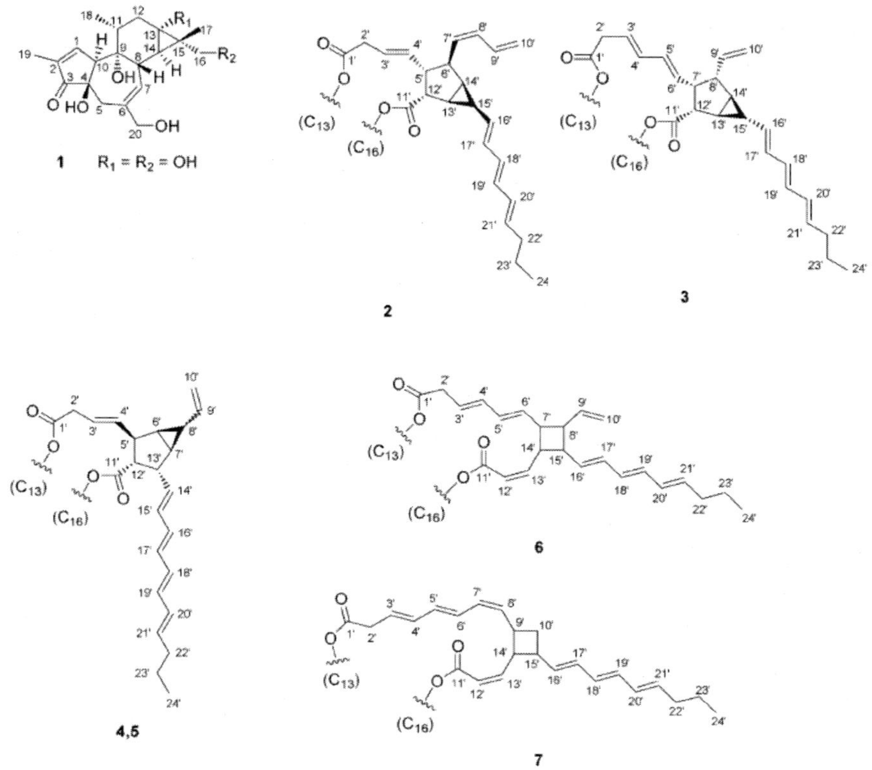

Source: Hass et al. (2002).

Figure 5. Different chemical structures of phorbol esters named DHPB.

Phorbol esters in *Jatropha* seed oil have six forms that are the isomers of chemical structures (Hass et al., 2002). They have a main structure of 12-deoxy-16-hydroxyphobol (Figure 5(1)) and also contain different side chains R_1 and R_2 to form six different isomers of phorbol esters (Figures 5(2-7)).

Figures 5(4) and 5(5) are actually epimer and could not be separated by chromatography technique. All of these phorbol esters structures are named as DHPB.

2.2.2. Physical and Chemical Properties of Phorbol Esters

2.2.2.1. Description

Phorbol esters are isolated as white crystals or powders. When isolated from volatile organic solvents (ether, methylene dichloride) during fractionation of oil, they form brittle foams which change to amorphous which are soften at temperature below 100°C. Phorbol-12-myristate 13-acetate (TPA) is like phorbol, strongly retains solvent molecules which it forms addition compounds. The same probably applies to other phorbol esters as well. They are soluble in water and polar organic solvents.

Anhydrous phorbol (crystallized from water) has a melting point of 250-251°C. Phorbol crystallized from ethanol and methanol retains solvent molecules tenaciously and these “alcohol phorbols” have sharp melting points in the region of 230-240°C.

2.2.2.2. Stability

Phorbol esters are very sensitive to acid, alkali, elevated temperatures, light and atmospheric oxygen. Solid TPA appears to be stable when stored in the dark at -20°C. It shows slow decomposition at 4°C within 3 months in the dark and more extensive decomposition at 25°C in diffuse daylight within 3 months. The solution of TPA in dimethyl sulfoxide may be kept at -20°C in the dark for 6 months. Solution of TPA in ethanol may be kept in the dark under nitrogen at -4°C in the dark for 5 months. At -4°C there are only traces of decomposition, while at 25°C (in acetone, ethyl acetate or methylene chloride) autoxidation is extensive. The main products have been identified and consist mainly of oxidation products at the double bonds (Schimdt and Hecker, 1975; Jacobson et al., 1975; Ohuchi and Levine, 1978).

2.2.2.3. Chemical Reactivity

Hecker and Schmidt (1974) review phorbol esters and its esters. Phorbol esters reduce Fehling’s and Tollen regents, and form esters and ethers. The C_5 carbonyl group shows weak activity in the reaction with carbonyl agents but is reduced by sodium borohydride. The double bonds are subjected to reduction and to autoxidation. The primary alcohol group at C_{20} is oxidized to the aldehyde with MnO_2 or CrO_3.

2.2.2.4. Biological of Phorbol Esters

The phorbols themselves do not induce tumors but promote tumor growth following exposure to a subcarcinogenic dose of a carcinogen. They are rapidly absorbed through the skin and probably the intestinal tract. They may cause severe irritation of tissues (skin, eyes, mucous membranes and lungs) and induce sensitivity. Laboratory operations should be conducted in a fume hood and glove. If phorbol esters contact skin, wash with soap and cold water, avoid washing with solvents.

Highly irritant factors to skin are isolated from the seed oil of four *Jatropha* species (Adolf et al., 1984). These irritant factors are determined and that one is new polyunsaturated esters of 12-deoxy-16-hydroxyphorbol. The seed oil of *J. curcas* L. in Thailand is intended to produce in large amounts for the use as a substitute of a biodiesel and an ingredient in commercial printing ink. The irritant factors are tumor promoters, therefore its widely use might result in exposure of a large population to tumor promoters. In 1987 the irritant factors were partially purified from the seed oil of *J. curcas* L. in Thailand (Horiuchi et al., 1987). It shows the tumor-promoting activity in 12-Deoxy-16-hydroxyphorbol-4'-[12' ,14'-butadienyl]-6'-[16',18',20'-nonatrienyl]-bicyclo [3.1.0] hexane-(13-*0*)-2'- [carboxylate] -(16-*0*)-3'-[8'-butenoic-10']ate (DHPB) and 12-*O*-tetradecanoylphorbol-13-acetate (TPA) when it is experimented on mouse skin. The results showed that DHPB (unsaturated acid) has slightly weaker biological effect than TPA (saturated acid). TPA is widely used as standard phorbol esters in biochemical experiment.

2.2.2.5. Experimentation on Phorbol Esters

Many researches study and try to detoxify the phorbol esters substances in oil of *J. curcas* L. as follows.

The demand and supply gap of vegetable oil in the world because of the oil price increasing. *J. curcas* L. as an energy crop and *J.* seed oil is produced for biodiesel. In 1996, Foidl et al. developed *J. curcas* L. They study a technical process to produce methyl ester and ethyl ester from seed oil. The results shows that the fuel properties of both esters are followed the standard properties of biodiesel. Shweta et al. (2005) illustrate that the combination of sonication and enzyme treatment with a commercial preparation of pH 9 ledds to 97% oil yield within 2 hours.

J. curcas L. has a large number of potential utilizations. The seed weighs about 0.75 g, contains 30-32% protein and 60-66% lipid (Liberalino et al., 1988) indicating a good nutritional value. However, the seed or oil is found to be toxic to mice (Adam, 1974), rat (Liberalino et al., 1988), calves, sheep and goats (Ahmed and Adam, 1979), human (Mampane et al., 1987) and chickens (Samia et al., 1992). Hence, it is restricted to use as a food or feed source. The biological effects of phorbol esters are found by Aitken et al. in 1986. The biological effects are tumor promotion, cell proliferation, activation of blood platelets, lymphocyte mitogenesis, inflammation, prostaglandin production and stimulation of degranulation in neutrophils.

So, phorbol esters substances are interested and in 1988 Mitsuru et al. find a new type of phorbol esters which has a macrocyclicdicarboxylic acid diester structure. It is isolated from the seed oil of *J. curcas* L. and its structure is proposed as an intramolecular 13, 16 diester of 12-deoxy-16-hydroxyphorbol-4'-[12',14'-butadienyl] -6'- [16',18',20'-nonatrienyl] -bicyclo [3.1.0] hexane-(13-*0*) -2'-[carb-oxylate] -(16-*0*)-3'- [8'-butenoic-10'] ate (DHPB). The results show that DHPB is tumor promotion with weaker biochemical activity than 12-*o*-tetradecanoylphorbol-13-acetate (TPA).

In 1995, Gandhi et al. provide data on toxicity of *Jatropha* seed oil which contains phorbol esters. A toxic fraction of the phorbol esters is isolated from the oil and LD_{50} is tested in rats. The acute oral LD_{50} of the oil is 6 mg/kg body weight in rats. Gross et al. (1997) suggest a method for detoxification of oil by extraction phorbol esters using ethanol. This method is in economic effort because of a lot of solvent consumption.

The toxic of phorbol esters substances have different biochemical activities depending on species of *J. curcas* L. In 1997, Makkar et al. evaluated the non-toxic and toxic varieties of *J. curcas* L. They describe that *Jatropha* meal contains high protein, high energy and low fiber. The amino acids composition of meals from the non-toxic and toxic varieties is also similar. The meal contains significant level of trypsin inhibitor, lectin and phytate. Their levels do not differ much between the non-toxic and toxic varieties. The differences between non-toxic and toxic varieties are the amount of phorbol esters content. The amount of phorbol esters in non-toxic from Mexico is 0.11 mg/g of kernel whilst toxic varieties content about 3.45 mg/g of kernel.

The biological effects of phorbol esters are necessary to find routes for detoxification of the oil. In 2000, Hass et al. experiment the edible oil processing steps on phorbol esters detoxification. They find that deacidification step and bleaching step are efficient for phorbol esters removal by 55% whereas degumming step and odor removal step are not effective on phorbol esters removing. In the same year, Rug and Ruppel (2000) also find phorbol esters to be an effective biopesticide against diverse fresh-water snails. Extracts from *J. curcas* L. are found to be toxic against snails transmitting *Schistosomamansoni* and S. *haematobium*. When compared with aqueous extract, methanol extract shows the highest toxicity against all organisms that are tested with values 25 ppm for cercariae and the snail *Biomphalariaglabrata* and 1 ppm for the snails *Bulinustruncates* and B. *natalensis*. Attenuation of cercariae leading to reduced infectivity in mice could be achieved in concentration below those exporting acute toxicity.

Jatropha oil or methanol extract of *Jatropha* oil containing phorbol esters has also been shown to have strong insecticidal effects against *Busseolafusca* and *Sesamiacalamistis* larvae (Mengual, 1997) and pesticidal effects against *Sitophiluszeamays* and *Callosobruchuschinesis* and deterred their oviposition on sprayed corn and mungbeans seeds (Solsoloy and Solsoloy, 1997).

2.3. Adsorption

Deacidification and bleaching steps of the traditional refinery oil process can reduce phorbol esters content in seed oil of *J. curcas* L. up to 55% (Wilhelm et al., 2000). This research is interested to select the method of phorbol esters elimination in seed oil of *J. curcas* L. The bleaching agent can adsorb color of oil and may also adsorb phorbol esters.

Therefore, the adsorption process is selected a method to eliminate phorbol esters from seed oil.

3. Materials and Methods

3.1. Materials

3.1.1. *Jatropha curcas* Seed Oil from KU Biodiesel Project, Kasetsart University

3.1.2. *Jatropha curcas* Press Cake from KU Biodiesel Project, Kasetsart University

3.1.3. Reagents

- Methanol (Analytical grade, Merck, Germany)
- Acetronitrile (HPLC grade, Merck, Germany)
- Hexane (Analytical grade, Merck, Germany)
- Sodium chloride (Analytical grade, APS, Australia)
- Sodium hydroxide (Analytical grade, J.T. Baker, US)
- Potassium hydroxide (Analytical grade, J.T. Baker, US)
- Heptane (Analytical grade, Merck, Germany)
- Boron trifluoride in methanol (BF_3,14%v/v, Supelco Analytical, US)

3.1.4. Chemical Standards

- 4β, 9α, 12β, 13α, 20-pentahydroxytiglia-1, 6-dien-3-on-12β-myristate-13- acetate (tetradeca-noylphorbolacetate, TPA) (Sigma, US)
- Mehylheptadecanoate
- Fatty acid methyl esters mixture (C_8-C_{24}) (Supelco Analytical, US)

3.1.5. Adsorbent Agents

- Activated carbon from Patum Vegetable Oil Co., Ltd., Thailand
- Bentonite 150 mesh from Patum Vegetable Oil Co., Ltd., Thailand
- Bentonite 200 mesh from Patum Vegetable Oil Co., Ltd., Thailand
- Chitosan (Seafresh chitosan (lab), Thailand)
- Chitin (Seafresh chitosan (lab), Thailand)

3.2. Equipments

3.2.1. Balance 4 digit (Percisa, 120A, US)
3.2.2. Soxhlet Extraction Instrument (BÜchi, B811, Switzerland)
3.2.3 Gas Chromatography Instrument (Agilent Technique, 6890N, US)
3.2.4. High Performance Liquid Chromatography with UV detector (Shimadzu, LC-10AC, Japan)
3.2.5. High Performance Liquid Chromatography with diode array and mass spectrometry detector (Agilent Technique, US)
3.2.6. Surface area analysis (Quanta Chrome, Atosorpb-1)
3.2.7. Platfrom Shaker (Inonva 2100, Japan)
3.2.8. Centrifugation (Mermle, Z323, Germany)
3.2.9. Autoclaving (Dectra, US)
3.2.10. Rota evaporator (BÜCHI, R114, Switzerland)
3.2.11. Overhead Stirrer (Ingenieurbüro, CAT R17, Germany)
3.2.12. Hot air oven (Binder, German)
3.2.13. Fourier transform infrared spectrophotometer (Perkin Elmer System 2000, US)
3.2.14. Kjeldakl-digestion and distillation system (C. Gerhardt GmbH and Co. KG, VAP30, Germany)
3.2.15. Water bath (Memmert, WB14, Germany)

3.3. Methods

3.3.1. Elimination of Phorbol Esters from Seed Oil by Adsorption Process

3.3.1.1. Selection of the Most Suitable Adsorbent

About 25 ml of seed oil were mixed with 0.8 g of each adsorbent (activated carbon, bentonite150, bentonite200, chitin and chitosan) into a 250 ml Erlenmeyer flask. Adsorption was experimented at room temperature for 45 min of stirring time and 200 rpm of stirring rate. After that adsorbent and seed oil were separated by filtration with filter paper No.1. Extracted phorbol esters from 10 g of the seed oil with methanol. Content of phorbol esters was analyzed by HPLC. The best adsorbent was selected from maximum adsorbed phorbol esters from seed oil.

3.3.1.2. Optimization of the One-Time Adsorption

About 25 ml of seed oil were mixed with the most suitable adsorbent from experiment 3.3.1.1 in a 250 ml Erlenmeyer flask. The experiments were continued in order to find the optimum conditions of adsorption in terms of the following factors:

a. Amount of adsorbent: 0.4, 0.6, 0.8, 1.0, 1.2, 1.4 and 2.0 g.
b. Stirring time: 15, 30, 45, 60, 120 and 180 min.
c. Temperature: 32, 45, 65, 85 and 120°C.
d. Stirring rate: 0, 100, 150, 200, 250 and 300 rpm.

After each experiment, the adsorbent and seed oil were separated by filtration with filter paper no.1. Phorbol esters substance was extracted from the seed oil and the amount of phorbol esters was analyzed by HPLC.

3.3.1.3. Optimization of Two-Time Adsorption

About 25 ml of seed oil from the one-time adsorption were mixed with the most suitable adsorbent from experiment 1.1 in a 250 ml Erlenmeyer flask. The experiments were continued in order to find the optimum conditions of adsorption in terms of the following factors:

a Amount of adsorbent: 0.2, 0.4, 0.6, 0.8 and 1.0 g.
b Stirring time: 0, 15, 30 and 45 min.

After each experiment, the adsorbent and seed oil were separated by filtration with filter paper No.1. Phorbol esters substance was extracted from the seed oil and amount of phorbol esters was analyzed by HPLC.

3.4. Analytical Methods

3.4.1. Phorbol Esters Extraction

3.4.1.1. Phorbol Esters Extraction in *Jatropha* Seed Oil

Phorbol esters in seed oil were extracted from 10 g of seed oil with 10 ml of methanol for 4 times using funnel separation. The combined extracts were centrifuged to separate the extracts from the oil residue at 3000 rpm for 15 min. Then, the extracts were concentrated with rotaevaporator at 45°C and 200 mmHg. After that, the concentrated extracts were transferred into a 25 ml volumetric flask and filled up to 25 ml with methanol. The extracts were stored at - 20°C for HPLC analysis.

3.4.2. Analysis of Phorbol Esters Content

3.4.2.1. Preparation of Sample

About 1.5 ml of the extracts were filtered through 0.45 μl membrane prior to the measurement of the phorbol esters by HPLC.

Figure 6. Phorbol esters extraction from *Jatropha* seed oil with funnel separation.

The operation condition was 1 ml/min flow rate, 35°C thermal control column, 280 nm UV detector and 20 μl samples were injected. The mobile phase was acetronitrile and deionized water (80:20, v/v) with isocratic mode.

3.4.2.2. Calibration Curve of Phorbol Esters Standard

The standard tetradecanoylphorbolacetate (TPA) was dissolved in methanol. TPA standard concentrations were prepared at 10, 20, 30, 40 and 50 ppm, respectively. After that, phorbol esters content in the form of TPA was measured as previously described in 3.2.1. Area peak and phorbol esters concentration were plot on y-axis and x-axis, respectively. The calibration curve was a straight line that passed through the origin point. The external standard technique was used to quantify phorbol esters content according to the standard curve.

3.4.3. Conformation of Phorbol Esters by LC-MS/MS Using Multiple Reaction Monitoring (MRM) Mode

Chromatographic separation of phorbol of phorbol esters was preformed on C18 water Atlantis (5μm 2.7 × 50 mm). Isocratic program was used with mobile phase, consisted of solvent (50 mmol ammonium acetate + acetonitrile, 9 + 1 v/v). The flow rate was 0.2 ml/min, the injection volum was 40 μl MS/MS condition: MS/MS was performed on a Micromass Quattro Ultima triple-quadrapole spectrometer equipped with ESI source. The parameters used for the mass spectrometry under ESI^+ mode were as follows: capillary voltage 3.00 KV, cone voltage 50 V, source block temperature 120 °C, cone gas 52 l/h, desolvation temperature 350 °C, desolvation gas 593 l/h.

4. Results and Discussion

4.1. Raw Material

The seed oil was pressed from *Jatropha* seed by screw press and then the seed oil and the press cake were separated. After that, the seed oil was filtered through filter paper No.1. Phorbol esters content in seed oil analyzed by HPLC was approximately 3-6 mg/g. However, phorbol esters content of seed oil depended on the region culture of *J. curcas* L. For example, phorbol esters content of *Jatropha* varieties from Mexico was about 0.11 mg/g while phorbol esters content of *Jatropha* varieties from Thailand contained about 3-6 mg/g.

The adsorbents used in this study were activated carbon, bentonite 150, bentonite 200, chitin and chitosan as presented in Figure 7. Activated carbon was a general term covering carbon material mostly derived from charcoal. Bentonite was special clay and usually formed from weathering of volcanic ash. Bentonite 150 and bentonite 200 are the same material with different particle sizes. The number '150' and '200' behind the word 'bentonite' present the mesh bentonite particle size in mesh. Chitin and chitosan are co-polymers of carbohydrates and included the derivative of Nitrogen-Glucose combination cation molecules. Chitin is a natural organic compound which is insoluble in water and general organic solvents but dissolved in concentrate organic acids. Chitosan can dissolve in various organic acids and form gel, granule and fiber and is used in surface coating. We can find the hard-shelled of shellfish which have many profits for plants, animals and humanity like these.

The Physical Properties of Adsorbents

The physical properties of 5 adsorbents in this experiment were presented in Table 1. Among all adsorbents, the activated carbon has the highest surface area and is strong alkaline which the activated carbon has 923.80 m^2/g and pH equal of 9.84, respectively. Particle sizes of bentonite 150 and 200 were 150 and 200 mesh, respectively. Bentonite 150 and 200 are strong acidic condition which pH of 3.05 and 2.50, respectively.

Figure 7. The adsorbents of Activated carbon (a), Bentonite 150 (b), Bentonite 200 (c), Chitin (d) and Chitosan (e).

Table 1. The physical properties of 5 adsorbents

Types of adsorbent	Particle size (mesh)	Surface area (m^2/g)	Pore volume (cc/g)	Pore size (°A)	pH
Activated carbon	150	923.80	0.4818	60.1060	9.84
Bentonite 150	150	190.40	0.0885	101.1500	3.05
Bentonite 200	200	327.30	0.1488	101.4000	2.50
Chitin	40	1.13	$5.856E^{-4}$	83.8200	5.60
Chitosan	60	-	-	-	7.90

Figure 8.Comparison of *Jatropha* seed oil before and after adsorption with adsorbents.

Chitin and chitosan have large particle sizes with 40 and 60 mesh, respectively, indicating that they contain low surface area and are neutral. However, the surface area of chitosan could not be detected because the temperature of surface area test was 300°C where the chitosan cannot stand for. Figure 8 showed the *Jatropha* seed oil after the adsorption experiment. It demonstrated that all adsorbents improved the clarity of *Jatropha* seed oil. However, bentonite 150 and 200 showed the best adsorption capability as indicated by the clearest of *Jatropha* seed oil after adsorption, followed by activated carbon, chitin and chitosan. The highest adsorption capability of bentonite could be because bentonite is usually applied as a bleaching agent in a traditional edible oil refining.

4.2. Phorbol Esters Chromatogram

Analysis of phorbol esters by an isocratic mixture of 80% acetronitrile and 20% deionized water showed the retention time of phorbol esters about 8-12 min as referred with the method of Wink et al. (1997). Within 8-12 min, the phorbol esters chromatogram contained 5 peaks and therefore the total area of the 5 phorbol esters peaks were used for quantification (Figure 9).

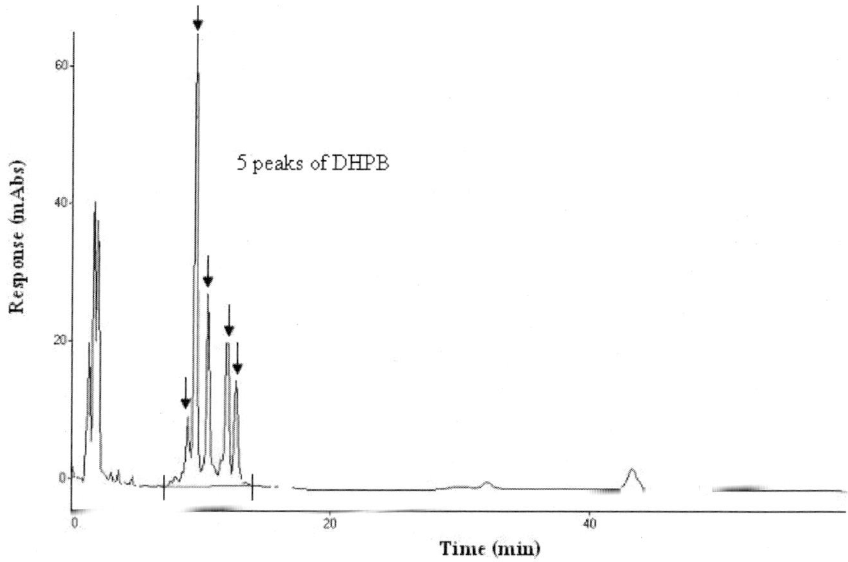

Figure 9. Chromatogram of phorbol esters (DHPB) in *Jatropha* seed oil.

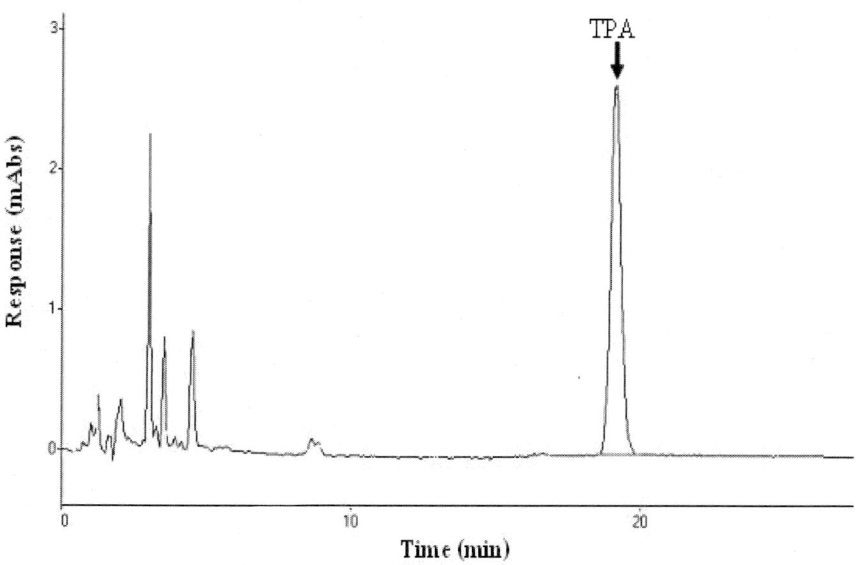

Figure 10. Chromatogram of phorbol-12-myristate-13-acetate (TPA) standard.

Although the phorbol esters found in *Jatropha* seed was DHPB (12-Deoxy-16-hydroxyphorbol-4'-[12',14'-butadienyl]-6'-[16',18',20'-nonatrienyl]-bicyclo[3.1.0] hexane-(13-*0*)-2'-[carboxylate]-(16-*0*)-3'-[8'-butenoic-10']ate, Hass and Mittelbach, 2000) but it is not commercially available. TPA was used as external standard for quantification of phorbol esters according to Wink et al. (1997). However, this could lead to far higher values than when using DHPB in this experiment, only the relative decrease of phorbol esters was interesting thus this difference was neglected (Gläser, 1991).

The chromatograms of standard phorbol esters (phorbol-12-myristate 13-acetate, TPA), phorbol esters obtained from *J. curcas* oil and *Jatropha* wood were presented in Figures 10.

4.3. Elimination of Phorbol Esters from *Jatropha* Seed Oil

4.3.1. Selection of the Most Suitable Adsorbent

The phorbol esters adsorption by different adsorbents was presented in Figure 11.

All experiments were performed under the same condition which was 3.2% adsorbents (w/w) at room temperature with 200 rpm stirring rate. When increased the stirring time of adsorption, the phorbol esters adsorption was also increased. The results illustrated that the highest % phorbol esters obtained from activated carbon, bentonite 150, bentonite 200, chitosan and chitin were 18.01% (15 min stirring time), 96.09% (45 min stirring time), 98.38% (45 min stirring time), 8.12% (300 min stirring time) and 12.28% (300 min stirring time), respectively.

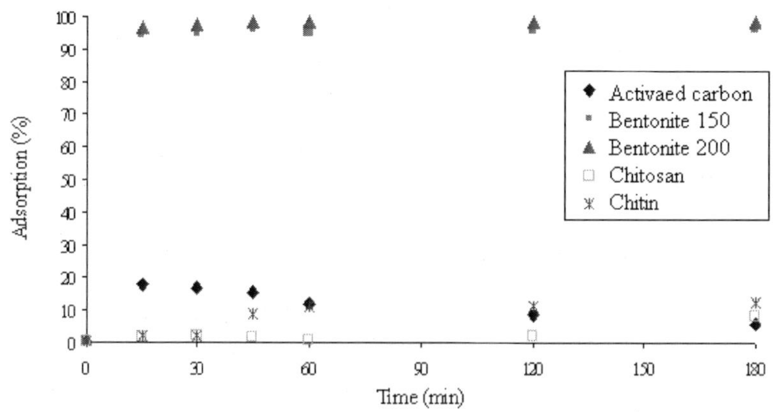

Figure 11. Phorbol esters adsorption capability in *Jatropha* seed oil by 5 adsorbents.

When considered the physical properties of adsorbents, pH and pore size, the results indicated that pH of adsorbent affected the adsorption capability more than their surface areas and pore sizes. It demonstrated that bentonite 150 and 200 had the adsorption capability more than those of activated carbon, chitin and chitosan. This might be due to stronger acidity of bentonite 150 and 200 (pH 3.05 and 2.50, respectively) compared with the basicity of activated carbon and chitosan.

Therefore, it resulted in the hydrolysis reaction simultaneously with the adsorption of phorbol esters. The comparison between bentonite 150 and 200 showed that bentonite 200 had smaller size, stronger acidity and more contact surfaces area and therefore showed higher efficiency for phorbol esters adsorption. As shown that the adsorption capability was higher under the acidic condition. When bentonite 150 and 200 were applied, they contained the same pore size (101°A).

Therefore, these results indicated that pH affected of adsorbent the capability more than the surface area and pore size of adsorbents. The phorbol esters adsorption of bentonite 200 reached equilibrium after 15 min and also reached the highest adsorption capability at 96.72%. The removal of phorbol esters from our study was far better than that obtained from Hass et al. (2000) which phorbol esters were removed only by 55%.

As a result, bentonite 200 was the most suitable adsorbent for phorbol esters adsorption from the *Jatropha* seed oil and applied for further experiment.

4.3.2. Optimization of One-Time Adsorption

The optimum phorbol esters adsorption condition by bentonite 200 (the most suitable adsorbent from previous section) was summarized as follows. The stirring time, amount of bentonite 200, temperature and stirring rate of adsorption were optimized. The results were showed in Figure 12.

Figure 12(a) showed the effect of stirring time on adsorption. The optimum stirring time was 15 min where the phorbol esters adsorption was 96.72%. When increasing the stirring time, % adsorption was also increased until at 45 min where the adsorption became flatten out. In Figure 12 (b), the results indicated an increase in adsorption capability with increasing the bentonite amount with the maximum adsorption at 99.63% when 2.0 g of bentonite 200 were applied. However, the optimum adsorption with 0.8 g bentonite 200 was selected because at the amount of bentonite 200 more than 1.0 g, the adsorption became flattened out.

The effect of temperature on adsorption was shown in Figure 12(c).

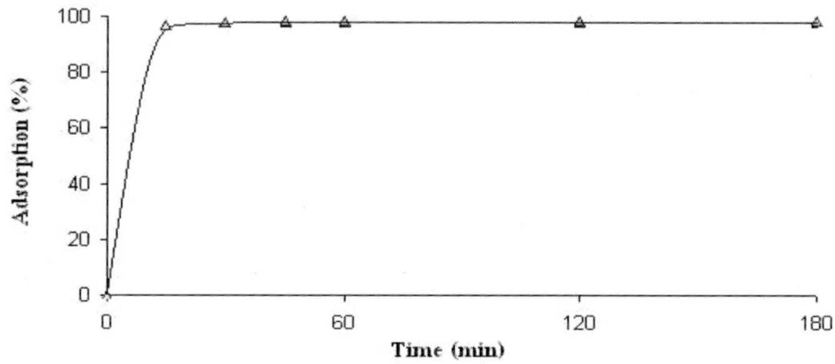

a

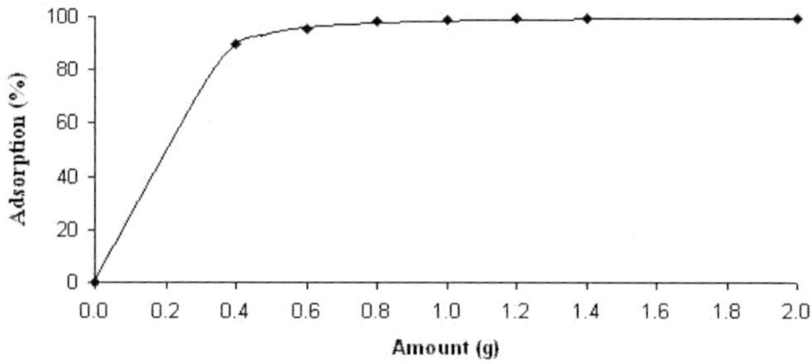

b

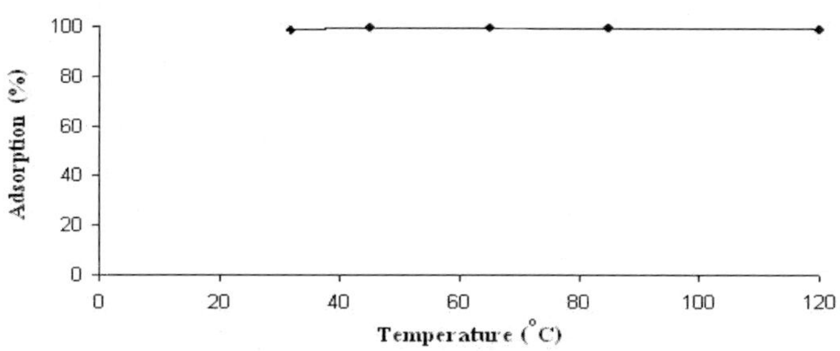

c

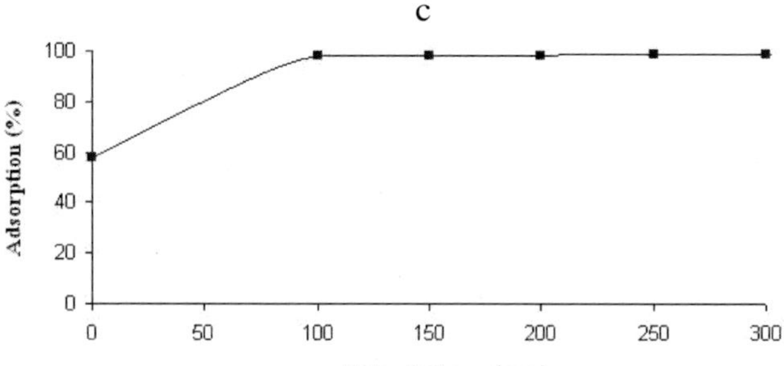

d

Figure 12. The effect of stirring time (a), amount of bentonite 200 (b), temperature (c) and stirring rate (d) on one-time adsorption of phorbol esters in *Jatropha* seed oil.

The adsorption capability of all tested temperatures was about 99 % with remaining phorbol esters content as small as 0.11 mg/g (equivalent to that in the non-toxic of *Jatropha curcas* L.). The effect of stirring rate on adsorption capability of bentonite 200 was shown in Figure 12(d). As the stirring rate increased, the adsorption capability also increased. However, the stirring rate faster than 100 rpm gave constant adsorption about 98%, therefore, 100 rpm was selected as the most optimum stirring rate for adsorption by bentonite 200.

The result also showed that the equilibrium condition of adsorption was obtained under the condition: 3.2% (w/v) bentonite 200 at room temperature, 100 rpm of stirring rate and stirring time for 15 min, illustrating high phorbol esters adsorption efficiency at 96.72%, 98.45%, 98.25% and 98.38%, respectively. Temperature and stirring rate of adsorption almost did not affect the phorbol esters adsorption with bentonite 200, whereas stirring time and amount of bentonite 200 highly affected the adsorption.

4.3.3. Optimization of the Two-Time Adsorption

Even though one-time adsorption of *Jatropha* seed oil with bentonite 200 showed high efficiency of phorbol esters removal up to 98.44% or 0.0928 mg/g phorbol esters remained in *Jatropha* seed oil, it would be of more advantageous if there was no remaining phorbol esters at all. As a result, the two-time adsorption was experimented when the seed oil after the one-time adsorption was selected to the adsorption again with new bentonite 200 under the new adsorption condition. Stirring rate and temperature of the two-time adsorption were fixed at 100 rpm and 32°C, respectively, according to the previous results. The results were demonstrated in Tables 2 and 3.

Tables 2 and 3 showed the effect of bentonite 200 amount and stirring time on the second time adsorption, respectively. Phorbol esters content in *Jatropha* seed oil was 5.9670 mg/g. After one-time of adsorption, the remaining phorbol esters in *Jatropha* seed oil was 0.0928 mg/g.

The results remonstrated that an increase in bentonite 200 amount increased a little of adsorption capability with the remaining phorbol esters content in *Jatropha* seed oil about 0.0213 mg/g or 99.64% of adsorption. Increasing of stirring rate also increased a little of adsorption capability with the remaining phorbol esters content in *Jatropha* seed oil about 0.0216 mg/g or 99.64% of adsorption. Thus, it was concluded that the maximum adsorption efficiency of phorbol esters from the second time was about 99.60% with remaining phorbol esters in the seed oil of approximately 0.02 mg/g.

Figure 13 summarized the adsorption efficiency of bentonite 200 on phorbol esters for one and two times.

Table 2.The effect of bentonite 200 amount on the two-time adsorption

Weight of BT200 (g)	PEs content (mg/g)		Adsorption for one time (%)	Weight of BT200 (g)	PEs content (mg/g)	Adsorption for two-time (%)
	Before adsorption	After one-time adsorption				
0.8	5.9670	0.0928	98.44	0.0	0.0928	98.45
				0.2	0.0213	99.64
				0.4	0.0186	99.689
				0.6	0.0204	99.66
				0.8	0.0215	99.64
				1.0	0.0212	99.64

Table 3. The effect of stirring time on the two-time adsorption

Weight of BT200 (g)	PEs content (mg/g)		Adsorption for one time (%)	Weight of BT200 (g)	PEs content (mg/g)	Adsorption for two-time (%)
	Before adsorption	After one-time adsorption				
0.8	5.9670	0.0928	98.44	0	0.0928	98.44
				15	0.0216	99.64
				30	0.0213	99.64
				45	0.0174	99.71

PEs = Phorbol esters.
BT200 = Bentonite200.

The equilibrium condition was 0.2 g bentonite 200 (0.8%, w/v), 15 min stirring time at 32°C (room temperature) and 100 rpm stirring rate. The adsorption of phorbol esters was up to 99%.

Figure 14, the comparison between one and two-time adsorption showed the phorbol esters content remained in *Jatropha* seed oil and percentage of adsorption. The two-time adsorption could increase adsorption about 1.20% from the one-time adsorption and the remaining phorbol esters were about 0.0213 mg/g. When increased amount of bentonite 200 more than 0.8% (w/v) and increased stirring time longer than 15 min, it showed almost no effect on the adsorption capability. As a result, bentonite 200 had limited to adsorb phorbol esters in *Jatropha* seed oil in the two-time adsorption.

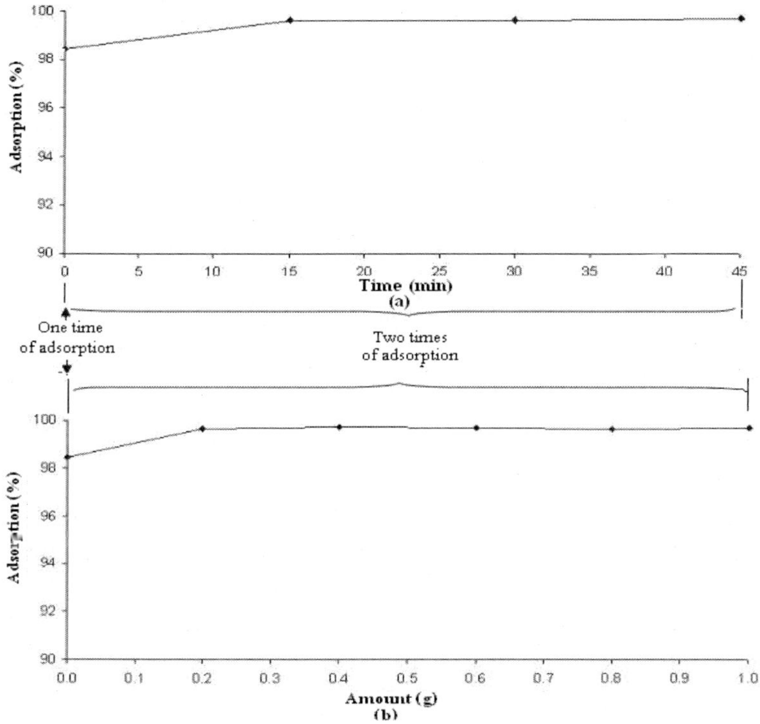

Figure 13. The effect of stirring rate (a) and bentonite 200 amount (b) on the two-time adsorption capability of phorbol esters.

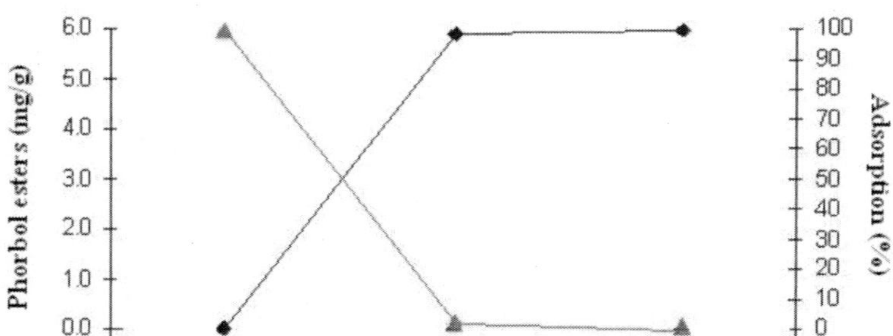

Time of adsorption	0	1st	2nd
Phorbol esters (mg/g)	5.9670	0.0928	0.0213
Adsorption (%)	0.0000	98.4448	99.6430

Figure 14. One-time and two-time of adsorption capability of phorbol esters in seed oil.

4.4. Confirmation of Phorbol Ester by LC-MS/MS

The confirmation of phorbol ester by LC-MS/MS with MRM mode of *Jatropha curcas* oil before adsorption showed two ionization peaks. The first one ionization peak represented parent molecule with mass 711 ionized to precursor ion with mass 311 and the second ionization peak ionized from precursor ion to produce an ion with mass 293.

This result could be explained assuming that phorbol ester fragmented by eliminating its ester groups (C_{13} and C_{16} of figure 2) and alcohol group (C_{20} of figure 2) to diterpene ester of tigliane type (molecular formula = $C_{20}H_{23}O_{23}$) resulting in precursor molecule with mass 311.

The skeleton was further fragmented by losing H_2O (molecular mass = 18) to produce ion with mass 293. Hence this characteristic pattern could be used to establish specific detection of phorbol ester residuce in *Jartropha curcas* oil after adsorption.

In the case of oil after adsorption. The result from HPLC chromatogram revealed a small peak occurring between 8-12 min (the amount of phorbol ester about 0.02 mg/g) but after confirmation it did not show the two ionization peaks, indicating that phorbol esters were not left in the oil after adsorption. In addition, the concentration of residue phorbol esters in oil after adsorption is lower than 0.11 mg/g phorbol esters that reported by Makkar and Becker (1997) in the nontoxic Mexican varieties, too.

CONCLUSION

Bentonite 200 was the most suitable adsorbent for phorbol esters adsorption from seed oil when compared among the activated carbon, bentonite 150, chitin and chitosan. The optimum condition was 15 min adsorption time, 3.2% (w/v) bentonite 200, 32°C temperature and 100 rpm stirring rate with maximum removal up to 98.00% or 0.09 mg/g phorbol esters remained in seed oil.

The 2nd adsorption showed the optimum condition at 0.8% (w/v) bentonite 200, 15 min stirring time at 32°C temperature and 100 rpm stirring rate with maximum removal up to 99.50% or 0.02 mg/g phorbol esters remained in seed oil. Liquid chromatogram-tandem mass spectrometry (LC-MS/MS) with multiple reaction monitoring (MRM) mode is useful technique to confirm phorbol esters left after adsorption by bentonite 200.

The results of our study show no two ionization peaks appear that indicate the adsorption technique by bentonite 200 is useful technique for phorbol esters elimination from *J. curcas* oil.

REFERENCES

[1] Adam, E. I. 1974. Toxic effects of *Jatropha curcas* in mice.*Toxicology*. 2: 67-76.

[2] Adam, E. I. 1979. Toxicity of *Jatroph acurcas* for goats.*Toxicology*. 4: 347-354.

[3] Adolf, W., Opferkuch, H. J., Hecker, E. 1984. Irritant phorbol derivates from four *Jatropha* species.*Phytochemistry*. 23 (1): 129-132.

[4] Aitken, A. 1986. In: *naturally occurring phorbol esters*. CRC Press, Boca Raton, FL. 271-288.

[5] Becker, K., Makkar, H. P. S. 1998. Effect of phorbol esters in carp (*Cyprinuscarpio* L.).*Veterinary and Human Toxicology*. 40 (2): 82-86.

[6] Evan, F. J. 1986. *Environmental hazards of diterpene esters from plants*. CRC Press, Boca Raton, FL. 1-31.

[7] Gandhi, V. M., Cheriaw, K. M., Mulky, M. J. 1995. Toxicological studies on Ratanjot Oil. *Food Chemistry Toxicology*. 33 (1): 39-42.

[8] Gläser, S. 1991. Untersuchung zu einem möglichen Gesundheits - und Krebsrisiko durch pflanzliche Arzneimittel sowie industriell genutzte Rohstoffe aus Euphorbiaceen-Quantitative Bestimmung von irritierenden und tumor-promovierenden Diterpenestern und Evaluierung durch biochemische biologische Tests. *Ph.D. thesis*, University of Heidelberg, Heidelberg.

[9] Gonzalez-Guerrico, A. M., Kazanietz1, M. G. 2005. Phorbol ester-induced apoptosis in prostate cancer cells via autocrine activation of the extrinsic apoptotic cascade.*Journal of Biological Chemistry*. 280: 38982-38991.

[10] Gross, H., Foidl, G., Foidl, N. 1997. Detoxification of *Jatropha curcas* press cake and oil and feeding experiments on fish and mice. *Biofuels and Industrial Products from Jatropha curcas Dbv*, Graz university. 179-182.

[11] Gübitz, G. M., Mittelbach, M., Trabi, M. 1999. Exploitation of the topical oil seed plant *Jatropha curcas* L. *Bioresource Technology*. 67: 73-82.

[12] Hass, W., Mittelbach, M. 2000. Detoxification experiments with seed oil from *Jatropha curcas* L. *Industrials Crops and Products*. 12: 111-118.

[13] Hass, W., Sterk, H., Mittelbach, M. 2002. Novel 12-deoxy-16-hydroxyphorbol diesters isolated from the seed oil of *Jatropha curcas*. *Journal of Natural Products*. 65: 1434-1440.

[14] Hecker, E., Schmidt, R. 1974. Phorbol esters: The irritants and cocarcinogens of cotton*tiglium* L. *Fortschritte der Chemie Organischer Naturstoffe*. 31: 377-467.

[15] Heller, J. 1996. *Physic nut. Jatrophacurcas L. promoting the conservation and use of underutilized and neglected crop*. Institute of Plant Genetics and Crop Plant Research, Gatersleben / International Plant Genetic Resources Institute, Rome.

[16] Horiuchi, T., Fujiki, H., Hirota, M., Suttajit, M., Suganuma, M., Yoshioka, A., Wongchai, V., Hecker, E., Sujimura, T. 1987. Precence of tumor promoters in the seed oil of *Jatrophacurcas* L. from Thailand.*Japan of Journal Cancer Research*. 78: 223-226.

[17] Jacobson, K., Wenner, C. E., Femp, G., Papahadjopoulous, D. 1975. Surface properties of phorbol esters and their interaction with lipid monolayers and bilayers.*Cancer research*.35: 2991-2995.

[18] Liberalino, A. A., Bambirra, E. A., Moraes-Santos, T., Viera, E. C. 1988. *Jatrophacurcas* L. seed: chemical analysis and toxicity. *Arquivos de biologia e technologia*. 31: 539-550.

[19] Liu, S. Y., Sporer, M., Jourdance, J., Henning, R., Li, Y. L., Ruppel, A. 1997. Anthraquinones in *Rheum palmatum* and *Rumexdentaus* (Polygonaceae) and phorbol esters from *Jatropha curcas* (Euphorbiaceae) with molluscicidal activity against the schistosomias vector snails *Oncomelania*, *Biomphalaria* and *Bulinus*. *Journal of Tropical Medicine and International Health*. 2: 179-188.

[20] Makkar, H. P. S., Aderibigbe, A. O., Becker, K. 1999. Comparative evaluation of non-toxic and toxic varieties of *Jatropha curcas* for chemical composition, digestibility, protein degradability and toxic factors.*Food Chemistry*. 62 (2): 207-215.

[21] Makkar, H. P. S., Becker, K. 1997. Potential of *J. curcas* seed meal as a protein supplement to livestock feed, constraints to its utilization and possible strategies to overcome constraints. *Biofuels and Industrial Products from Jatrophacurcas*. Dbv, Graz. 190-205.

[22] Makkar, H. P. S., Becker, K., Sporer, F., Wink, M. 1997. Study on nutritive potential and toxic constituents of difference provenances of *Jatropha curcas* L. *Agriculture Food Chemistry*. 45: 3152-3157.

[23] Mampane, K. J., Joubert, P. H., Hay, I. T. 1987. *Jatropha curcas*: use as a traditional Tswana medicine and its role as a cause of acute poisoning. *Phytotherapy Research*. 1: 50-51.

[24] Mengual, I. 1997. *Extraction of bioactive substances from J. curcas L. and bioassays on Zonocerus variegates, Sesamiacalamistis and Busseolafuscafusca for characterization of insectidal properties*. Dbv, Graz University.211-215.

[25] Mitsuru, H., Maitree, S., Hiroko, S., Yasoyoki, E., Koichi, S., Vichai, W., Erich, H., Hirota, F. 1988. A New tumor promoter from the seed oil of *Jatrophacurcas* L., an intramolecular diester of 12-Deoxy-16-hydroxyphobol.*Cancer Research*. 48: 5800-5804.

[26] Nwosu, M. O., Okafor, J. I. 1995. Preliminary studies of the antifungal actives of some medicinal plants against *Basidiobolus* and some other pathogenic fungi.*Mycoses*. 38: 191-195.

[27] Ohuchi, K., Levine, L. 1978. Stimulation of prostaglandin synthesis by tumor promoting phorbol-12, 13 diesters in canine kidney (MDCK) cells.*J. Biol. Chem*. 253: 4783-4790.

[28] Rug, M., Ruppel, A. 2000. Toxic activities of the plant *Jatropha curcas* against intermiate snail hosts and larvae of schitosomes.*Tropica Medicine and International Health*. 5: 423-430.

[29] Samia, M. A., Badwi, E. L., Mausa, H. M., Adam, S. E. I. 1992. Response of brown chicks to low level of *Jatropha curcas*, Riciouscomnunis or their mixture.*Veterinary and Human Toxicology*. 34: 304-306.

[30] Schmidt, R., Hecker, E. 1975. Autoxidation phorbol esters under normal storage conditions.*Cancer Reserch*. 35: 1375-1377.

[31] Shweta, S., Aparna, S., Gupta, M. N. 2005. Extraction of oil from *Jatropha curcas* L. seed kernels by combination of ultrasonication and aqueous enzymatic extraction.*Bioresearch Technology*. 96: 121-123.

[32] Solsoloy, A. O., Solsoloy, T. S. 1997. Pesticidal efficacy of formulated *J. curcas* oil on pets of selected field crops.*Biofuels and Industrial Products from Jatropha Curcas*.Dbv, Graz University.216-226.

[33] Wilhelm, H. S., Martin, M. 2000. Detoxification experiments with the seed oil from *Jatropha curcas* L. *Industrial Crops and Product*. 12: 111-118.

[34] Wink, M., Koschmieder, C., Sauerwein, M., Sporer, F. 1997. Phorbol esters of *Jatropha curcas*-biological activities and potential applications.*Biofuels and Industrial Products from Jatropha curcas*.Dbv, Graz University.10: 160-166.

In: Seed Oil
Editor: Alexis Varnham
ISBN: 978-1-63463-056-6

Chapter 7

SESAME OIL AND SESAMOL AS PROTECTIVE AND THERAPEUTIC AGENTS AGAINST DRUG-INDUCED SINUSOIDAL OBSTRUCTION SYNDROME

Srinivasan Periasamy and Ming-Yie Liu*
Department of Environmental and Occupational Health,
National Cheng Kung University, College of Medicine,
Tainan, Taiwan

ABSTRACT

Sinusoidal obstruction syndrome (SOS), previously known as veno-occlusive disease (VOD), occurs in patients undergoing hematopoietic cell transplantation and chemotherapy. SOS is historically called Gulran disease in Afghanistan and senecio disease in South Africa; it dates back to 1920. Pyrrolizidine alkaloids (PAs) in herbal preparations such as tea and Chinese medicine induce SOS. PAs in grasses and animal feed cause acute and chronic poisoning in cattle. The chemotherapeutic drugs oxaliplatin and cyclophosphamide also cause SOS. The search for a novel and effective therapy for chemotherapeutic-drug-induced-SOS continues. Sesame oil is a nutrient-rich antioxidant popular in alternative medicine and traditional health foods in Asian countries. Sesame oil and its lignan sesamol have been proved effective for treating various drug-induced and

* myliu@mail.ncku.edu.tw.

chemically induced liver injuries. Sesame oil and sesamol maintain glutathione and reduce myeloperoxidase activity, nitrate content, lipid peroxidation (LPO), and the recruitment of inflammatory cells in SOS. In addition, they downregulate matrix metalloproteinase (MMP)-9 expression and upregulate tissue inhibitor of metalloproteinases (TIMP)-1, laminin, and collagen in SOS. We hypothesize that sesame oil and sesamol would be useful for treating PA-mimicking chemotherapeutic drug-associated SOS.

HISTORY

Hepatic veno-occlusive disease (VOD) was historically called Gulran disease in Afghanistan because it has consistently occurred in the Gulran district of Herat Province in western Afghanistan. The history of VOD dates back to 1920. It was first reported in 1920 as 80 cases of *senecio disease*, in the town of George, Western Cape Province, South Africa, *bread poisoning* caused by *Senecio ilicifolius* and *S. burchelli*, which grow as weeds in the wheat fields. The chief symptoms are abdominal pain and vomiting with ascites (Wilmot and Robertson, 1920). In Uzbekistan, about 1500 cases, called *camel belly*, were reported to have occurred in 1931 and 1945 (Dubrovinskii, 1946). The largest outbreak ever reported in the Gulran District of western Afghanistan was from 1974 to 1976; it affected an estimated 7800 people and caused 1600 deaths. Gulran disease was attributed to eating bread made from wheat contaminated with the seeds of a weed, locally called *charmac*, which includes *Heliotropium popovii* (*H. popovii Riedl subsp. gillianum Riedl*) which contains pyrrolizidine alkaloids (PAs), primarily heliotrine, and the liver biopsy led to a diagnosis of hepatic veno-occlusive disease (Tandon and Tandon, 1975; Tandon et al., 1978). A second outbreak occurred from 1999 to 2001 with an estimated 400 cases and over 100 deaths (Bower, 2001), and in February 2008, Afghanistan's outbreak and disease surveillance system responded to rumors of another outbreak of Gulran disease and identified 38 cases of massive ascites and four deaths that appeared to be associated with eating contaminated wheat flour (Kakar et al., 2010).

Veno-Occlusive Disease and Sinusoidal Obstruction Syndrome

VOD as a diverse clinical entity was first described in South Africa and was associated to the ingestion of PAs contained in Senecio tea (Willmot and

Robertson, 1920). The characteristic occlusion of the terminal venules of the liver by blood in the individuals who drank this tea led to the term veno-occlusive disease. Several species of plants containing PAs can cause VOD and have been associated with epidemics of this disease in developing nations (Datta et al., 1978). VOD developing after allogeneic stem cell transplantation was first described in 1979 (Berk et al., 1979), and stem cell transplantation is the most common cause of VOD in the Western Hemisphere (McDonald et al., 1984).

Sinusoidal changes are primary events in the pathology of the disease. The term sinusoidal obstruction syndrome (SOS) may describe the condition better than does VOD (DeLeve et al., 2002). Hepatocytes have a canalicular surface, and they form the bile canaliculi and a basolateral sinusoidal surface. The sinusoidal surface is lined with a single layer of endothelial cells, the fenestrae of which allow free communication between the sinusoids and the extravascular space of Disse. A delicate network of collagen fibers supports the sinusoidal lining. Endothelial injury seems to be the initiating event in the cascade of events that lead to hepatic changes and the clinical manifestation of SOS. Experimental studies in a rat model of SOS contributed to the understanding of events that lead to these changes (DeLeve et al., 1999).

Pyrrolizidine Alkaloids in Herbal Preparations: Tea- and Chinese Medicine-Induced SOS

The potential hepatotoxicity of herbal preparations and other botanicals has been underestimated because of a common misconception by the public that they are harmless. They are often used for self-medication without supervision. Several species of PA-containing plants can cause SOS and have been associated with epidemics in developing countries such as Jamaica, India, Egypt, Iraq, and South Africa (Tandon et al., 1976; Moayad and Muntaha, 1998). Exposure to PAs has been regarded as one of the two major causes of SOS (Fu et al., 2004). The earliest case of PA-induced SOS was reported in 1920 and associated with drinking PA-containing herbal tea (Willmot and Robertson, 1920). Since then, about 8160 PA-poisoning cases have been documented in many countries, including Afghanistan, Britain, China, Germany, Hong Kong, India, Jamaica, South Africa, Switzerland, and the United States (Dai et al., 2007). More problematically, many unknown PA-containing plants, which have similar appearances and names and often cannot be distinguished from one another, are possibly misused as folk medicines and foods. Because of the severe adverse impact of PA-induced SOS on public health, many regulations have been created to restrict or prohibit using PA-containing medicinal herbs (Lin et al., 2011).

The principal medicinal genera in current use are *Senecio, Borago, Lithospermum, Heliotropium*, and *Eupatorium*. Toxic reactions to some of those medicinal herbs have been reported in people who have ingested them for many months. PAs became a herbal medicine safety concern about 20 years ago, when an enthusiast who drank comfrey (*Symphytum officinale*) tea and took six comfrey-pepsin tablets daily for 6 months developed VOD (Stickel and Seitz, 2000). Comfrey, a well-known medicinal herb characterized by U.S. FDA researchers as having been "one of the most popular herb teas in the world", contains PAs that are capable of causing liver damage (Betz et al., 1994).

Pyrrolizidine Alkaloids in Chinese Herbs

The commonly used Chinese herbs that are currently known to contain PA are these: *Zicao*, which is obtained from *Lithospermum erythrorhizon* and *Arnebia euchroma* of the Boraginaceae; *Kuandonghua*, which comes from *Tussilago farfara* of the family *Asteraceae*. In Japan, *Petasites japonicus* has been used as *kuandonghua*; *Peilan*, from *Eupatorium fortunei* and *E. japonicum* of the family *Asteraceae* (Tang, 1995); and *Qianliguang*, obtained from *Senecio scandens* of the family *Asteraceae*. In addition, *Senecio scandens* and tablets from its extract were officially listed in the Chinese Pharmacopoeia 1977, which indicated that it was for bacterial diarrhea, enteritis, conjunctivitis, and respiratory tract infections. It is still used in composition formulas to treat rhinitis, ulcerative colitis, and burns. It contains two PAs senecionine and seneciphylline. Two species of Chinese herbs *Eupatorium* (*E. japonicum* and *E. cannabinum*) contain PAs, and a species of *Crotalaria* (*C. assamica*) was found to contain one PA, monocrotaline (MCT) (Edgar et al., 1992). Other Chinese herbs with PAs include various species of *Senecio* used in folk medicine (e.g., *S. argunensis* and *S. integrifolius*), *Emilia sonchifolia* (*Asteraceae*), and *Crotalaria sessiliflora. Heliotropium indicum* (*Boraginaceae*) is a folk remedy used in Taiwan to treat lung diseases and sore throat (Osungunna and Adedeji, 2011).

Herbal teas, including traditional *bush* teas (Fragoso-Serrano et al., 2012), contain PAs, traditional Chinese medicines, and dietary supplements (Fu et al., 2002); they also cause PA toxicity. A frequent complication in the use of traditional Chinese medicines is liver toxicity (Pittler and Ernst, 2003). Human exposure to PAs can occur through a number of routes, including herbal preparations and teas (Wiedenfeld, 2011), cereals and grains (Kakar et al., 2011), honey, food supplements, and salad leaves (Edgar et al., 2011; Kempf et al., 2011), and even milk (Hoogenboom et al., 2011). Pak et al. (2004) reported that articles which link liver toxicity to herbal preparations are escalating. The German Federal Institute for Risk Assessment (BfR), which analyzed 221 commercially available teas, herbal teas, and drugs as part of a project to determine PAs in food and feed, that more research is necessary to detect PAs in herbal teas before they are marketed and in those already on the market (BfR, 2013).

Acute and Chronic Pyrrolizidine Alkaloid Poisoning in Cattle

Approximately 600 species of *Crotalaria* (family *Fabaceae*; rattle pods) grow in tropical and subtropical regions of the world (Holland, 2002). Several species of *Crotalaria* are sources of the hepatotoxic PA MCT (Cheeke, 1998). All livestock species are susceptible to PA toxicity (EFSA, 2007). PAs sporadically poison grazing horses, cattle, sheep, pigs, and poultry whose feed grain is contaminated with their seeds (Bull et al., 1968). The acute poisoning and death of cattle occur because of a large intake of *Crotalaria* in a short period under conditions of natural grazing. Acute poisoning is characterized by extensive damage, necrosis, and hemorrhage of the liver (TRS, 2001; Copple et al., 2002). PAs are monoesters and diesters of the pyrrolizidine necine bases retronecine and otonecine from a variety of plant species, with 1,2-unsaturation of the necine base required for hepatotoxicity (Elias et al., 2011). In general, modern management of feeds and livestock herds has reduced the risk of PA toxicity considerably, but occasional intoxications are still reported. PA plant toxicity continues to be recorded in various parts of the world in the veterinary and other scientific literature (TRS, 2001). Nonetheless, *Crotalaria* poisoning is underreported (Nuhu et al., 2009). Reports of poisoning in livestock, together with the results of studies in experimental animals, suggest that species differ in their susceptibility to PAs. In general, sheep, goats, and rabbits appear to be more resistant and tolerate higher PA doses. In sheep, this tolerance is thought to be because of demonstrated detoxification by PA-destroying rumen microbes (Radostits et al., 2000). Rabbits and guinea pigs also appear less sensitive to PAs (McLean, 1970), while horses, pigs, and poultry are considered more sensitive (WHO-

IPCS, 1988). Various diet supplement treatments are ineffective against PA intoxication in livestock. Poisoned animals with clinical signs rarely recover; therefore, prevention is the best control measure (Stegelmeier et al., 2009). Veterinary teams usually use local remedies for *Crotalaria* poisoning, for example, peanut oil, atropine sulfate, and antidiarrheal agents; however, one experimental study (Srinivasan and Liu, 2012) shows that sesame oil is equally effective in treating acute MCT poisoning.

Pyrrolizidine Alkaloids: Metabolism and Action

PAs are minimally toxic in their original form. PAs are metabolized in the liver through a CYP (cytochrome P450) 3A-mediated transformation to N-oxides and conjugated dienoic pyrroles. Pyrroles are alkylating compounds that are highly reactive with proteins and nucleic acids. The complex of pyrroles with proteins and nucleic acids may persist in tissue and generate chronic injury, whereas N-oxides may be transformed into epoxides and toxic necines (Yang et al., 2001). CYP3A inducers increase PA toxicity, and CYP3A inhibitors decrease it. PAs decrease glutathione (GSH) in sinusoidal endothelial cells, which increases oxidative stress; oxidative stress is important in PA-induced SOS. GSH conjugates with dehydromonocrotaline to form a compound of much lower toxicity that is released in high concentration into bile (Wang et al., 2000). The increased oxidative stress can also affect collagen $\alpha 1$ transcription directly or by activating hepatic stellate cells, which ultimately leads to SOS (Chojkier et al., 1998). Moreover, PAs inhibit the proliferation of hepatocytes, decrease the levels of the anti-apoptotic protein Bcl-x, and increase the expression of the pro-apoptotic protein Bax, which ultimately leads to the release of cytochrome C from mitochondria and activates the intrinsic apoptotic pathway (Chojkier et al., 1998).

Monocrotaline Metabolism

MCT an alkaloid pyrrolizidine phytotoxin, is well-documented for its hepatic and cardiopulmonary toxicity in animals, including ruminants, and humans (Nobre et al., 2004). MCT toxicity requires cytochrome P-450-mediated bioactivation to reactive pyrrolic metabolite dehydromonocrotaline (Schultze and Roth, 1998). Dehydromonocrotaline is unstable and can continue through several metabolic pathways: (i) hydrolysis to 6,7-dihydro-7-

hydroxy-1-hydroxymethyl-5H-pyirrolizine (DHP), one of the major active metabolites; (ii) conjugation with GSH in the liver to form 7-enantiomers glutathionyl-6,7-dihydro-1-hydroxymethyl-5H-pirrolizine (7-GS-DHP) and 7,9-DP-diGSH; (iii) nucleophilic alkylation of cellular macromolecules, which is a process that highlights the toxic activity of dehydromonocrotaline; or (iv) release by circulation (Wang et al., 2005). Ester group hydrolysis by, for example, carboxylesterase, and then excretion of the acid and amino alcohol products, and the formation and excretion of highly water-soluble N-oxides are detoxification mechanisms. Dehydromonocrotaline, despite having a half-life of only a few seconds in aqueous media (Mattocks et al., 1990), is a powerful alkylating agent that binds to cellular DNA and proteins (Wagner et al., 1993; Lamé et al., 2005). Dehydromonocrotaline inhibits NADH-dehydrogenase activity in mitochondria, an effect associated with significantly reduced ATP synthesis (Mingatto et al., 2007).

Monocrotaline-Induced Sinusoidal Obstruction Syndrome

SOS is drug-induced liver injury that occurs in patients undergoing hematopoietic cell transplantation and chemotherapy with oxaliplatin. The dietary ingestion of pyrrolizidine alkaloid-MCT and cytotoxic drugs, e.g., azathioprine, cyclophosphamide, and busulfan, is a well-known major cause of SOS (Deleve, 2007). Clinicopathologic (Shulman et al., 1994) and experimental (DeLeve et al., 1999; Wang et al., 2000) studies suggest that the essential change in SOS occurs in the hepatic sinusoid. MCT-induced SOS initiates morphological change in the liver by rounding up sinusoidal endothelial cells (SECs) within 12 hours. Between 24 and 48 hours, blood penetrates the sinusoidal lining beneath the rounded-up SECs and dissects the sinusoidal lining from the space of Disse. The sinusoid is obstructed by an embolism of SECs ultimately denuded of their sinusoidal lining (DeLeve et al., 2003).

Oxaliplatin-Induced SOS in Patients with Cancer

Oxaliplatin, the newest platinum derivative in standard chemotherapy, differs from cisplatin in that the amine groups of cisplatin are replaced by diaminocyclohexane. Its full chemical name, oxalate (trans-l-1,2-diaminocyclohexane) platinum, refers to the presence of an oxalate "leaving group" and the diaminocyclohexane carrier ligand, which are responsible for its unique properties. Oxaliplatin in plasma rapidly undergoes non-enzymatic transformation into reactive compounds by displacing the oxalate group, after which diaminocyclohexane platinum complexes enter the cell and cause cytotoxicity. Oxaliplatin's cytotoxicity comes from DNA damage that arrests DNA synthesis, inhibits RNA synthesis, and triggers immunologic reactions (Alcindor and Beauger, 2011).

Oxaliplatin induces SOS in non-tumor-bearing hepatic parenchyma (Rubbia-Brandt et al., 2004). Patients treated with oxaliplatin develop SOS associated with higher morbidity and prolonged hospital stays (Nakano et al., 2008). Clinically, SOS is characterized by hyperbilirubinemia (> 2 mg/dl), ascites, hepatomegaly, hepatic sinusoidal dilatation, hepatocyte atrophy, perisinusoidal fibrosis, nodular regenerative hyperplasia, portal hypertension, and weight gain (Rubbia-Brandt et al., 2004, 2010). SOS is caused by the toxic effect of oxaliplatin on SECs (Rubbia-Brandt et al., 2004). The ensuing swelling of SECs and loss of sinusoidal wall integrity impairs sinusoidal blood flow causes congestive obstruction (DeLeve, 2007), which eventually leads to peliosis, centrilobular hepatic vein fibrotic obstruction, perisinusoidal fibrosis, and nodular regenerative hyperplasia (Rubbia-Brandt et al., 2010).

Beneficial Effects of Sesame Oil and Sesamol

Sesame oil is consumed in various forms by humans worldwide. Sesame oil, a component of traditional health foods in various Asian countries, protects against atherosclerosis, hypertension, and aging. Sesame oil contains sesamolin, sesamin, and sesamol (392, 238, and 11.5-16.1 mg/l00 g of oil, respectively) (Mohamed and Awatif, 1998). Sesamol is formed by the thermal hydrolysis of sesamolin. Sesame seed contains two lignans: sesamin and sesamolin. When sesame seeds are roasted, their sesamolin is converted to sesamol (Fukuda et al., 1986). Sesamol (5-hydroxy-1,3-benzodioxole 3,4-[methylenedioxy]phenol) a powerful antioxidant, scavenges singlet oxygen (Kim et al., 2003). Sesamol has a phenolic and a benzodioxole group in its molecular structure. The phenolic groups of molecules are responsible for its antioxidant activity (McPhail et al., 2003). Sesamol inhibits lipid peroxidation, hydroxyl radical-induced deoxyribose degradation, and DNA cleavage (Joshi et al., 2005). In Swiss albino mice, scavenging the free radicals, activating the endogenous antioxidant enzymes, protecting the hematopoietic system, and preventing DNA damage are likely to be the mechanisms for the radioprotective activity of sesamol (Parihar et al., 2006). Recent studies report that sesamol attenuates experimental epilepsy (Hassanzadeh et al., 2014), inflammatory bowel disorder (Kondamudi et al., 2014), cardiomyopathy (Chennuru et al., 2013), Parkinson's disease (Sonia Angeline et al., 2013), and stress-related mucosal disease (Hsu et al., 2013).

Sesame Oil and Sesamol Protect Against Monocrotaline-Induced SOS

Prophylactic sesame oil attenuates MCT-induced SOS. Sesame oil decreases hepatic injury and the expression and activity of MMP-9, but it increases TIMP-1 expression in MCT-induced SOS. Prophylactic sesame oil's inhibition of the activity and expression of MMP-9 and its antioxidant activity are responsible for its protection against experimental SOS. Prophylactic sesame oil mitigates the breakdown and loss of the extracellular matrix proteins laminin and collagen, both of which preclude necrosis and the collapse of hepatocyte cytoskeletons. Sesame oil attenuates SOS by inhibiting oxidative stress, myeloperoxidase activity, lipid peroxidation, and nitrate content, and by maintaining glutathione levels, all of which protect against SOS (Periasamy et al., 2013b). Sesame oil attenuates acute MCT poisoning in a rat model. Both sesame oil and peanut oil attenuate pancreatic, lung, and liver injury in acute monocrotaline poisoning. Sesame oil efficiently decreases steatosis, but peanut oil does not. Therefore, sesame oil is more beneficial than peanut oil in attenuating multiple organ injury in acute MCT poisoning.

Therapeutic sesamol attenuates MCT-induced SOS. Sesamol's anti-MMP-9 property may be responsible for its protection against experimental SOS. MMP-9 is a typical coagulator important for the breakdown and necrosis of hepatocytes. Inhibiting the release of active MMP-9 attenuates the severity of SOS. Sesamol inhibits the release of active MMP-9, inhibits MMP-9's activity, and upregulates TIMP-1. Sesamol mitigates the breakdown and loss of the extracellular matrix proteins laminin and collagen, which precludes the collapse of the hepatocyte cytoskeleton and prevents necrosis. Therapeutic sesamol reduces hemorrhage around the central, hepatic, and portal veins in MCT-induced SOS. Pathological analysis shows no rounding up of SECs, and the architecture of the hepatocytes appears to be normal. Sesamol reduces tissue injury by inhibiting the recruitment of inflammatory cells and inhibiting the expression and activity of MMP-9, thereby attenuating the breakdown of the cytoskeleton proteins of hepatocytes after the onset of SOS (Periasamy et al., 2011).

Therapeutic intervention of subcutaneous sesamol attenuates liver injury and improves metabolic function in MCT-induced SOS in rats. However, therapeutic oral sesame oil is ineffective against MCT-induced SOS in rats; in fact, it may increase the burden to the damaged liver after SOS has developed. These findings warn against using sesame oil as a therapeutic medication or a nutritional supplement to treat the accidental dietary ingestion of PAs and the cytotoxic effects of cyclophosphamide and busulfan, all of which may cause SOS in patients (Periasamy et al., 2013a).

Conclusion

Sesame oil and sesamol show no observable adverse effects when they are used to treat MCT intoxication in rats. They protect against SOS by downregulating MMP-9 expression, upregulating TIMP-1 expression, and inhibiting oxidative stress in rats. They may attenuate liver injury and improve metabolic function in PA-induced and chemotherapeutic regimen-induced SOS in humans. However, their efficacies in humans are yet to be tested.

References

Alcindor, T; Beauger, N. Oxaliplatin: a review in the era of molecularly targeted therapy. *Current Oncology*, 2011 18, 18–25.

Berk, PD; Popper, H; Krueger, GR; Decter, J; Herzig, G; Graw, RG Jr. Veno-occlusive disease of the liver after allogeneic bone marrow transplantation: possible association with graft-versus-host disease. *Annals of Internal Medicine*, 1979 90, 158–164.

Betz, JM; Eppley, RM; Taylor, WC; Andrzejewski, D. Determination of pyrrolizidine alkaloids in commercial comfrey products (Symphytum sp.). *Journal of Pharmaceutical Sciences*, 1994 83, 649–653.

BfR. Bundesamt für Risiokobewertung. Pyrrolizidine alkaloids in herbal teas and teas; Stellungnahme Nr. 018/2013 des BfR vom 5 July 2013 [Internet]. [cited 2013 Sep 3]. Available from: http://www.bfr.bund.de/cm/349/pyrrolizidine-alkaloids-inherbal-teas-and-teas.pdf 2013.

Bower, H. Drought Brings Back Gulran Disease, WHO Health Talks, Kabul, Afghanistan, 2001.

Bull, LB; Culvenor, CCJ; Dick, AT. The Pyrrolizidine Alkaloids. Their Chemistry, Pathogenicity and Other Biological Properties; *Frontiers of Biology Series 9*. Amsterdam, The Netherlands: North-Holland; 1968.

Cheeke, PR. *Natural Toxicants in Feeds, Forages, and Poisonous Plants*, 2nd ed. Danville, IL: Interstate., 1998, 338–352.

Chennuru, A; Saleem, MT. Antioxidant, lipid lowering, and membrane stabilization effect of sesamol against doxorubicin-induced cardiomyopathy in experimental rats. *Biomedical Research International*, 2013 2013, 934239.

Chojkier M; Houglum K; Lee, KS; Buck, M. Long and short term D-alpha-tocopherol supplementation inhibits liver collagen alpha (1) gene expression. *American Journal of Physiolology,* 1998 275, 1480–1485.

Copple, BL; Banes, A; Ganey, PE; Roth, RA. Endothelial cell injury and fibrin deposition in rat liver after monocrotaline exposure. *Toxicological Sciences,* 2002 65, 309–318.

Dai, N; Yu, YC; Ren, TH; Wu, JG; Jiang, Y; Shen, LG; Zhang, J. Gynura root induces hepatic veno-occlusive disease: a case report and review of the literature. *World Journal of Gastroenterology*, 2007 13, 1628–1631.

Datta, DV; Khuroo, MS; Mattocks, AR; Aikat, BK; Chhuttani, PN. Herbal medicines and veno-occlusive disease in India. *Postgraduate Medical Journal,* 1978 54, 511–515.

DeLeve, LD. Hepatic microvasculature in liver injury. *Seminars in Liver Disease*, 2007 27, 390–400.

DeLeve, LD; Ito, I; Bethea, NW; McCuskey, MK; Wang, X; McCuskey, RS. Embolization by sinusoidal lining cell obstructs the microcirculation in rat sinusoidal obstruction syndrome (hepatic venooclusive disease). *American Journal of Physiology - Gastrointestinal and Liver Physiology*, 2003 284, G1045–G1052.

DeLeve, LD; McCuskey, RS; Wang, X; Hu, L; McCuskey, MK; Epstein, RB; Kanel, G. Characterization of a reproducible rat model of hepatic veno-occlusive disease. *Hepatology,* 1999 29, 1779–1791.

DeLeve, LD; Shulman, HM; McDonald, GB. Toxic injury to hepatic sinusoids: sinusoidal obstruction syndrome (veno-occlusive disease). *Seminars Liver Disease*, 2002 22, 27–42.

Dubrovinskii, SB. [About the alimentary toxicosis caused by heliotrope.] [In Russian] *Journal Sovereign Protection and Health*, 1946 6, 17–21.

Edgar, JA; Colegate, SM; Boppré, M; Molyneux, RJ. Pyrrolizidine alkaloids in food: a spectrum of potential health consequences. *Food Additives & Contaminants: Part A,* 2011 28, 308–324.

Edgar, JA; Lin, HJ; Kumana, CR; Ng, MMT. Pyrrolizidine alkaloid composition of three Chinese medicinal herbs. *American Journal of Chinese Medicine,* 1992 20, 281–288.

EFSA (European Food Safety Authority), 2007. Opinion of the scientific panel on contaminants in the food chain on a request from the European commission related to pyrrolizidine alkaloids as undesirable substances in animal feeds. *The EFSA Journal*, 447, 1–51.

Elias, F; Latorre, AO; Pı´pole, F; Haraguchi, M; Górniak, SL; Hueza, IM. Haematological and immunological effects of repeated dose exposure of rats to integerrimine N-oxide from Senecio brasiliensis. *Food and Chemical Toxicology*, 2011 49, 2313–2319.

Fragoso-Serrano, M; Figueroa-González, G; Castro-Carranza, E; Hernández-Solis, F; Linares, E; Bye, R; Pereda-Miranda, R. Profiling of alkaloids and eremophilanes in miracle tea (Packera candidissima and P. bellidifolia) Products. *Journal of Natural Products*, 2012 75, 890–895.

Fu, PP; Xia, QS; Lin, G; Chou, MW. Pyrrolizidine alkaloids-genotoxicity, metabolism enzymes, metabolic activation, and mechanisms. Drug Metabolism and Reviews, 2004 36, 1–55.

Fu, PP; Yang, YC; Xia, QS; Chou, MW; Cui, YY; Lin, G. Pyrrolizidine alkaloids – tumorigenic components in Chinese herbal medicines and dietary supplements. *Journal of Food and Drug Analysis*, 2002 10, 198–211.

Fukuda, Y; Nagata, M; Osawa, T; Namiki, M. Chemical aspects of the antioxidative activity of roasted sesame seed oil and the effect of using the oil for frying. *Agricultural and Biological Chemistry,* 1986 50, 857–862.

Hassanzadeh, P; Arbabi, E; Rostami, F. The ameliorative effects of sesamol against seizures, cognitive impairment and oxidative stress in the experimental model of epilepsy. *Iran Journal of Basic Medical Sciences,* 2014 17, 100–107.

Holland, AE. A review of Crotalaria L. (Fabaceae: Crotalarieae) in Australia. *Austrobaileya*, 2002 6, 293–324.

Hoogenboom, LAP; Mulder, PPJ; Zeilmaker, MJ; van den Top, HJ; Remmelink, GJ; Brandon, EFA; Klijnstra, M; Meijer, GAL; Schothorst, R; van Egmond, HP. Carry-over of pyrrolizidine alkaloids from feed to milk in dairy cows. *Food Additives & Contaminants: Part A*, 2011 28, 359–372.

Hsu, DZ; Chen, YW; Chu, PY; Periasamy, S; Liu, MY. Protective effect of 3,4-methylenedioxyphenol (sesamol) on stress-related mucosal disease in rats. *Biomedical Research International,* 2013 2013: 481827.

Joshi, R; Kumar, MS; Satyamoorthy, K; Unnikrisnan, MK; Mukherjee, T. Free radical reactions and antioxidant activities of sesamol: pulse radiolytic and biochemical studies *Journal of Agricultural and Food Chemistry*, 2005 53, 2696–2703.

Kakar, F; Akbarian, Z; Leslie, T; Mustafa, ML; Watson, J; van Egmond, HP; Omar, MF; Mofleh, J. An outbreak of hepatic veno-occlusive disease in western Afghanistan associated with exposure to wheat flour contaminated with pyrrolizidine alkaloids. *Journal of Toxicology*, 2010 2010, 313280.

Kakar, F; Akbarian, Z; Leslie, T; Mustafa, ML; Watson, J; van Egmond, HP; Omar, MF; Mofleh, J. An outbreak of hepatic veno-occlusive disease in western Afghanistan associated with exposure to wheat flour contaminated with pyrrolizidine alkaloids. *Journal of Toxicology*, 2011 139, 728–735.

Kempf, M; Wittig, M; Schönfeld, K; Cramer, L; Schreier, P; Beuerle, T. Pyrrolizidine alkaloids in food: downstream contamination in the food chain caused by honey and pollen. *Food Additives & Contaminants: Part A,* 2011 28, 325–331.

Kim, JY; Choi, DS; Jung, MY. Anti-photooxidative activity of sesamol in methyleneblue- and chlorophyll-sensitized photooxidation of oil. *Journal of Agricultural and Food Chemistry*, 2003 51, 3460–3465.

Kondamudi, PK; Kovelamudi, H; Mathew, G; Nayak, PG; Rao, MC; Shenoy, RR. Investigation of sesamol on myeloperoxidase and colon morphology in acetic acid-induced inflammatory bowel disorder in albino rats. *Scientific World Journal,* 2014 2014, 802701.

Lamé, MW; Jones, AD; Wilson, DW; Segall, HJ. Monocrotaline pyrrole targets proteins with and without cysteine residues in the cytosol and membranes of human pulmonary artery endothelial cells. *Proteomics,* 2005 5, 4398–4413.

Lin, G; Wang, JY; Li, N; Li, M; Gao, H; Ji, Y; Zhang, F; Wang, H; Zhou, Y; Ye, Y; Xu, HX; Zheng, J. Hepatic sinusoidal obstruction syndrome associated with consumption of Gynura segetum. *Journal of Hepatology,* 2011 54, 666–673.

Mattocks, AR; Legg, RF; Jukes, R. Trapping of short-lived electrophilic metabolites of pyrrolizidine alkaloids escaping from perfused rat liver. *Toxicology Letters*, 1990 54, 93–99.

McDonald, GB; Sharma, P; Matthews, DE; Shulman, HM; Thomas, ED. Venocclusive disease of the liver after bone marrow transplantation: diagnosis, incidence, and predisposing factors. *Hepatology*, 1984 4, 116–122.

McLean, EK. The toxic action of pyrrolizidine (Senecio) alkaloids. *Pharmacological Reviews,* 1970 22, 429–483.

McPhail, DB; Hartley, RC; Gardner, PT; Duthie, GG. Kinetic and stoichiometric assessments of the antioxidant activity of flavanoids by electron spin resonance spectroscopy. *Journal of Agricultural and Food Chemistry,* 2003 51, 1684–1690.

Mingatto, FE; Dorta, DJ; Santos, AB; Carvalho, I; Silva, CHTP; Silva, VB; Uyemura, SA; Santos, AC; Curti, C. Dehydromonocrotaline inhibits mitochondrial complex I. A potential mechanism accounting for hepatotoxicity of monocrotaline. *Toxicon,* 2007 50, 724–730.

Moayad, YA; Muntaha, H. An outbreak of veno-occlusive disease of the liver in northern Iraq. *Eastern Mediterranean Health Journal*, 1998 4, 142–148.

Mohamed, HMA; Awatif, II. The use of sesame oil unsaponifiable matter as a natural antioxidant. *Food Chemistry*, 1998 62, 269–276.

Nakano, H; Oussoultzoglou, E; Rosso, E; Casnedi, S; Chenard-Neu, MP; Dufour, P; Bachellier, P; Jaeck, D. Sinusoidal injury increases morbidity after major hepatectomy in patients with colorectal liver metastases receiving preoperative chemotherapy. *Annals of Surgery*, 2008 247, 118–124.

Nobre, VM; Riet-Correa, F; Barbosa Filho, JM; Dantas, AF; Tabosa, IM; Vasconcelos, JS. [Poisoning by Crotalaria retusa (Fabaceae) in Equidae in the semiarid region of Paraíba]. [In Portuguese] *Pesquisa Veterinária Brasileira,* 2004 24, 132–143.

Nuhu, H; Abdurrahman, E; Shok, M. Ethnomedical studies of crotalaria species found in Zaria, northern Nigeria. *Nigerian Journal of Pharmaceutical Sciences*, 2009, 8, 46–53.

Osungunna, MO; Adedeji, KA. Phytochemical and antimicrobial screening of methanol extract of Heliotropium indicum leaf. *Journal of Microbiology and Antimicrobials*, 2011 3, 213–216.

Pak, E; Esrason, KT; Wu, VH. Hepatotoxicity of herbal remedies: an emerging dilemma. *Progress in Transplantation*, 2004 14, 91–96.

Parihar, VK; Prabhakar, KR; Veerapur, VP; Kumar, MS; Reddy, YR; Joshi, R; Unnikrishnan, MK; Rao, CM. Effect of sesamol on radiation-induced cytotoxicity in Swiss albino mice. *Mutation Research*, 2006 611, 9–16.

Periasamy, S; Chien, SP; Liu, MY. Therapeutic oral sesame oil is ineffectual against monocrotaline-induced sinusoidal obstruction syndrome in rats. *Journal of Parenteral and Enteral Nutrition*, 2013a 37, 129–133.

Periasamy, S; Hsu, DZ; Chen, SY; Yang, SS; Chandrasekaran, VR; Liu, MY. Therapeutic sesamol attenuates monocrotaline-induced sinusoidal obstruction syndrome in rats by inhibiting matrix metalloproteinase-9. *Cell Biochemistry and Biophysics*, 2011 61, 327–336.

Periasamy, S; Yang, SS; Chen, SY; Chang, CC; Liu, MY. Prophylactic sesame oil attenuates sinusoidal obstruction syndrome by inhibiting matrix metalloproteinase-9 and oxidative stress. *Journal of Parenteral and Enteral Nutrition*, 2013b 37, 529–537.

Pittler, MH; Ernst, E. Systematic review: hepatotoxic events associated with herbal medicinal products. *Alimentary Pharmacology & Therapeutics*, 2003 18, 451–471.

Radostits, OM; Gay, CC; Blood, CC; Hinchliff, KW. 2000. Veterinary Medicine, *A textbook of the diseases of cattle, sheep, pigs, goats and horses*. 9th Edition. W.B. Saunders & Co., Philadelphia, 1661–1664.

Rubbia-Brandt, L; Audard, V; Sartoretti, P; Roth, A.D; Brezault, C; Le Charpentier, M. Severe hepatic sinusoidal obstruction associated with oxaliplatin-based chemotherapy in patients with metastatic colorectal cancer. *Annals of Oncology*, 2004 15, 460–466.

Rubbia-Brandt, L; Lauwers, GY; Wang, H; Majno, PE; Tanabe, K; Zhu, AX; Brezault, C; Soubrane, O; Abdalla, EK; Vauthey, JN; Mentha, G; Terris, B. Sinusoidal obstruction syndrome and nodular regenerative hyperplasia are frequent oxaliplatin-associated liver lesions and partially prevented by bevacizumab in patients with hepatic colorectal metastasis. *Histopathology*, 2010 56, 430–439.

Schultze, AE; Roth, RA. Chronic pulmonary hypertension-the monocrotaline model and involvement of the hemostatic system. *Journal of Toxicology and Environmental Health Part B: Critical Review*, 1998 1, 271–346.

Shulman, HM; Fisher, LB; Schoch, HG; Henne, KW; McDonald, GB. Venoocclusive disease of the liver after marrow transplantation: histological correlates of clinical signs and symptoms. *Hepatology*, 1994 19, 1171–1180.

Sonia Angeline, M; Sarkar, A; Anand, K; Ambasta, RK; Kumar, P. Sesamol and naringenin reverse the effect of rotenone-induced PD rat model. *Neuroscience*, 2013 254, 379–394.

Srinivasan, P; Liu, MY. Comparative potential therapeutic effect of sesame oil and peanut oil against acute monocrotaline (Crotalaria) poisoning in a rat model. *Journal of Veterinary Internal Medicine*, 2012 26, 491–499.

Stegelmeier, BL; Gardner, DR; Davis, TZ. Livestock poisoning with pyrrolizidine-alkaloid-containing plants (Senecio, Crotalaria, Cynoglossum, Amsinckia, Heliotropium, and Echium spp.). Rangelands, 2009, 31, 35–37.

Stickel, F; Seitz, HK. The efficacy and safety of comfrey. *Public Health and Nutrition,* 2000 3, 501–508.

Tandon, BN; Tandon, HD; Tandon, RK; Narndranathan, M; Joshi, YK. An epidemic of hepatic veno-occlusive disease in central India. *Lancet,* 1976 2, 271–272.

Tandon, HD; Tandon, BN. Epidemic of liver disease-Gulran district, Herat Province, Afghanistan, Assignment Report EM/AFG/OCD/001/RB, World Health Organization, Geneva, Switzerland, 1975.

Tandon, HD; Tandon, BN; Mattocks AR. An epidemic of veno-occlusive disease of the liver in Afghanistan. Pathologic features. *American Journal of Gastroenterology*, 1978 70, 607–613.

Tang, W. Mutagenic and carcinogenic constituents of Chinese herbal medicines. *Abstracts of Chinese Medicine,* 1995 6, 227–271.

TRS (Technical report series NO. 2). Pyrrolizidine Alkaloids in Food. A Toxicological Review and Risk Assessment. *Australia New Zealand Food Authority,* 2001, 6.

Wagner, JG; Petry, TW; Roth, RA. Characterization of monocrotaline pyrrole induced DNA cross-linking in pulmonary artery endothelium. *American Journal of Physiology-Lung Cellular and Molecular Physiology*, 1993 264, L517–L522.

Wang, XD; Kanel, GC; Deleve, LD. Support of sinusoidal endothelial cell glutathione prevents hepatic veno-occlusive disease in the rat. *Hepatology,* 2000 31, 428–434.

Wang, YP; Yan, J; Beger, RD; Fu, PP; Chou, MW. Metabolic activation of the tumorigenic pyrrolizidine alkaloid, monocrotaline, leading to DNA adduct formation in vivo. *Cancer Letter*, 2005 226, 27–35.

WHO-IPCS (World Health Organisation-International Programme on Chemical Safety), 1988. Pyrrolizidine alkaloids. Environmental Health Criteria 80. WHO, Geneva, 1-345. Available from http://www.inchem.org/documents/ehc/ehc/ehc080.htm

Wiedenfeld, H. Plants containing pyrrolizidine alkaloids: toxicity and problems. *Food Additives & Contaminants: Part A*, 2011 28, 282–292.

Willmot, FC; Robertson, GW. Senecio disease, or cirrhosis of the liver due to Senecio poisoning. *Lancet*, 1920 2, 848–849.

Yang, YC; Yan, J; Churchwell, M; Beger, R; Chan, P; Doerge, DR; Fu, PP; Chou, MW. Development of a ^{32}P-postlabeling/HPLC method for detection of dehydroretronecine-derived DNA adducts in vivo and in vitro. *Chemical Research in Toxicology*, 2001 14, 91–100.

In: Seed Oil
Editor: Alexis Varnham
ISBN: 978-1-63463-056-6

Chapter 8

SESAME OIL AS A POTENTIAL THERAPEUTIC AGENT AGAINST NUTRITIONAL STEATOHEPATITIS

Srinivasan Periasamy and Ming-Yie Liu*
Department of Environmental and Occupational Health, National Cheng Kung University, College of Medicine, Tainan, Taiwan

ABSTRACT

Nonalcoholic fatty liver disease (NAFLD) is highly prevalent in the general population. Nonalcoholic steatohepatitis (NASH), also called *nutritional steatohepatitis* and *nutritional fibrosing steatohepatitis*, can progress to liver failure and hepatocellular carcinoma.

Nutritional fibrosing steatohepatitis has been called "a tale of a two-hit hypothesis": the "First Hit" is characterized by hepatic injury and fat accumulation, and the "Second Hit" is characterized by hepatic oxidative stress, inflammation, and insulin resistance.

Managing NAFLD focuses particularly on diet and exercise; managing NASH focuses on lifestyle modifications, control of associated metabolic issues, and pharmacological therapy for liver injury.

Successful care and treatment require an integrative approach.

* Corresponding author: Ming-Yie Liu. Department of Environmental and Occupational Health, National Cheng Kung University, College of Medicine, 138 Sheng-Li Road, Tainan 70428, Taiwan. E-mail: myliu@mail.ncku.edu.tw.

Proposed pharmacological therapies for NASH include vitamin E, ursodeoxycholic acid (a drug used to dissolve gallstones), pioglitazone (one of a class of drugs called thiazolidinediones that are used to treat type 2 diabetes), and metformin (used to treat type 2 diabetes); however, drugs are therapeutically limited and may produce adverse effects. The search for a novel and effective medication to treat nutritional steatohepatitis continues. Sesame oil is nontoxic, antioxidant-rich, and nutritional oil, and it is effective against various diseases models; it attenuates both the first and second hits of nutritional steatohepatitis. Sesame oil attenuates hepatic injury and steatosis, reduces levels of triglycerides, nitric oxide, malondialdehyde (a biomarker of lipid peroxidation), tumor necrosis factor-α, interleukin-6, interleukin-1β, leptin, tissue growth factor-β1, α-smooth muscle actin, fibrosis, and the activity of matrix metalloproteinase-2 and -9, but it increases tissue inhibitor of metalloproteinases-1 and peroxisomal proliferator-activated receptor-γ expression. Thus, we hypothesize that sesame oil would be useful for treating NASH.

History of the Discovery of NAFLD

Ludwig et al. (1980) introduced the term "nonalcoholic steatohepatitis" (NASH) to describe the histologic findings of macrovesicular steatosis, hepatocyte necrosis, and fibrosis in those who did not drink alcohol.

Other names used to describe this clinical entity include "nonalcoholic steatonecrosis", "fatty liver hepatitis", and "nonalcoholic fatty hepatitis" (Sanyal and AGA, 2002). That macrovesicular hepatic steatosis is associated with inflammatory changes and fibrosis in obese patients has been known for several decades. Clinically, however, it was generally ignored.

It is impossible to differentiate between the hepatic histopathology in patients with nonalcoholic and alcoholic steatohepatitis when they both include macrovesicular steatosis, ballooning degeneration, hepatocyte necrosis, Mallory bodies, and fibrosis (Peters et al., 1975). Obese patients who had neither abused alcohol nor undergone weight-loss surgery, and patients with diabetes had similar hepatic lesions (Falchuk et al., 1980; Hornboll et al., 1982).

Definition and Classification of Non-Alcoholic Fatty Liver Disease

Non-alcoholic fatty liver disease (NAFLD) is defined as fatty liver (FL), i.e., an accumulation of lipids inside the hepatocytes exceeding 5% of the weight of the liver, without hepatitis B or C virus infection, and without "excessive" ethanol intake ($\geq$ 120 g/day) (Angulo and Lindor, 2002; Neuschwander-Tetri and Caldwell, 2003). Secondary causes of NAFLD are [a] *Genetic/metabolic diseases*: Wolman's disease, Wilson's disease, Weber-Christian disease, lypodystrophic diseases, hemochromatosis, hypobetalipoproteinemia; [b] *Infective*: hepatitis C virus, hepatitis B virus; [c] *Extra-hepatic conditions*: cardiac failure, irritable bowel disease (IBD), hypothyroidism, intestinal bacterial overgrowth syndrome, other neoplastic diseases, polycystic ovarian disease, pregnancy; [d] *Nutritional*: total parenteral nutrition, protein malnutrition, prolonged starvation, jejunoileal bypass, high carbohydrate diet, and [e] *Drugs*: nonsteroidal antiinflammatory drugs (NSAIDs), tetracycline, tamoxifen, chloroquine, corticosteroids, estrogens, calcium antagonists, amiodarone, perhexiline maleate.

NAFLD includes a wide range of liver abnormalities, ranging from steatosis to steatohepatitis and fibrosis (Angulo and Lindor, 2002; Neuschwander-Tetri and Caldwell, 2003). Although steatosis usually has a benign prognosis, steatohepatitis and fibrosis may develop into cirrhosis and hepatocarcinoma (Bellentani et al., 2010).

Pathogenesis and Risk Factors of NASH

NASH is the most severe histologic form of NAFLD. The diagnosis and staging of NASH with uniform criteria are still being debated. Insulin resistance is associated with obesity and is central to the pathogenesis of NAFLD (LaBrecque et al., 2014). Insulin resistance, an excess accumulation of free fatty acids (FFAs), and an increased intracellular formation and buildup of toxic lipid metabolites combine to provoke an inflammatory response that elicits the progression to NASH. The buildup of triglycerides in the hepatocytes is a marker for disturbed lipid metabolism and shows increased lipid trafficking (Cusi, 2009). NASH can remain asymptomatic for years, or can progress to cirrhosis and hepatocellular carcinoma. Important histopathological features of NASH are steatosis, hepatocellular ballooning, and lobular inflammation; fibrosis is not part of the histological definition of NASH. One general hypothesis for the pathogenesis of NASH is the "multi-hit hypothesis", with metabolic syndrome playing a main role. The characteristics of the multiple "hits" differ in patients and are mostly undefined at present (LaBrecque et al., 2014).

PROGNOSIS AND COMPLICATIONS OF NASH

NAFLD progresses to NASH to cirrhosis and, finally, to hepatocellular carcinoma. NAFLD does not aggravate hepatotoxicity or the adverse effects of pharmacological agents, including hydroxy-methylglutaryl coenzyme A (HMG-CoA) reductase inhibitors. NAFLD and comorbid obesity and related metabolic factors may aggravate other liver diseases. A liver biopsy may indicate the severity of the disease, but only fibrosis, and not inflammation or necrosis, has been established to predict a prognosis for NASH. Histological advancement to end-stage liver disease may occur: NASH with bridging fibrosis or cirrhosis. Hepatitis C or HIV comorbid with NAFLD degrades prognoses and makes the NAFLD immune to therapy. Hepatitis C is commonly concomitant with steatosis, which may complicate a diagnosis of hepatitis C vs. NASH vs. both together. End-stage NASH is an often under-recognized cause of NASH-related cirrhosis: progressive fibrosis may be obscured by stable or improving steatosis and serologic features, mainly in older patients with NASH. Cryptogenic cirrhosis increases the risk of hepatocellular carcinoma. The major causes of mortality in patients with NASH comorbid with cirrhosis are liver failure, cardiovascular disease, sepsis, variceal hemorrhage, and hepatocellular carcinoma (LaBrecque et al., 2014).

ANIMAL MODELS OF NASH

Few animal models replicate the pathology and chronicity of fibrosing human liver diseases. A chronic (17 weeks) high-fat methionine and choline deficient (MCD) diet characterizes the rodent model of hepatic fibrosis (George et al., 2003). This model induces a pattern of perivenous and pericellular hepatic fibrosis with steatosis and is pathologically associated with steatohepatitis; these features are similar to steatohepatitis in humans. In addition, they characterize hepatocytes as the site of lipid peroxidation that develops during the evolution of hepatic fibrosis. The sequence of pathogenic events during prolonged ingestion of the MCD diet involves early lipid accumulation and lipid peroxidation in hepatocytes, followed by liver cell injury (increased ALT) and inflammation, stellate cell activation, profibrotic gene upregulation in stellate cells, and, eventually, obvious hepatic fibrosis (George et al., 2003).

In fibrosing steatohepatitis, fibrosis begins in a centrizonal, pericellular location, with fibrotic strands surrounding lipid-loaded hepatocytes. These histopathological features are similar to those seen in disorders of lipid-associated hepatic fibrosis in humans, such as in alcoholic liver diseases and NASH. In contrast, the pathology differs from that in simple choline deficiency, in which steatosis is associated with progressive portal fibrosis rather than steatohepatitis (Murray et al., 1986). In the MCD diet model, steatosis, chronic hepatocyte injury, and inflammation are followed by several weeks fibrosis and the activation of stellate cells.

A similar sequence of events occurs in human NASH, despite possible differences between the factors that cause steatosis in mice and rats fed the MCD diet (Rinella and Green, 2002; George et al., 2003).

Oxidative stress and the resultant liberation of cytokines are the two interrelated processes that promote hepatic fibrogenesis in response to chronic liver injury. In MCD diet-fed rodents, steatosis is a result of the high dietary fat content as well as the methionine and choline deficiency. The choline deficiency leads to hepatic steatosis. In addition, the methionine deficiency reduces glutathione (GSH) biosynthesis; GSH depletion impairs antioxidant defenses against pro-oxidants (George et al., 2003). The MCD diet causes oxidative stress in murine models of steatohepatitis. Oxidative damage is also prominent in the liver of humans with NASH, (MacDonald et al., 2001; Seki et al., 2001), as well as in the histologically similar disorder of alcoholic hepatitis (Angulo and Lindor, 2002; Tilg and Diehl, 2000).

Dietary Modifications and Lifestyle Changes in NASH

Dietary modification, weight loss, and exercise are beneficial because they reduce insulin resistance and normalize NAFLD (Nobili et al., 2008). However, few studies (Huang et al., 2005; Nobili et al., 2008) have used biopsy results to evaluate histological improvement in NAFLD.

Studies on different diets have all reported the normalization of ALT levels 1-3 months after the start of dietary changes: [a] restriction of caloric intake to < 25 kcal/kg/d of ideal body weight, [b] a daily 600-800 calorie intake reduction, [c] restriction of caloric intake to < 30 kcal/kg/d (Vajro et al., 2008), [d] low-calorie/low-carbohydrate intake (40-45% of caloric intake) (Huang et al., 2005; Naniwadekar, 2010), [e] restriction of total dietary fat content to < 30% of the caloric intake with < 10% of the caloric intake from saturated fats (Vajro et al., 2008).

Studies on weight loss in adult patients with NASH (Huang et al., 2005) and pediatric patients with NAFLD (Nobili et al., 2008) have documented, with biopsies, histological attenuation of steatosis and inflammation.

Although studies comparing the efficacy of different types of diets to encourage weight loss have not been able to prove the superiority of one diet over another, none has looked at NAFLD as an endpoint (Dansinger et al., 2005; Naniwadekar, 2010).

TREATING NASH

Vitamin E and Antioxidants

That vitamin E decreases oxidative stress provides a rationale for its use in patients diagnosed with NASH. According to the two-hit hypothesis, the first hit involves accumulating excess fat in the liver cells because of insulin resistance, and that leads to hepatic steatosis.

The second hit involves oxidative stress that causes lipid peroxidation and activates inflammatory cytokines, which leads to NASH (Chitturi and Farrell, 2001; McCullough, 2002).

Many studies (Abdelmalek et al., 2009; Chitturi and Farrell, 2001; Hickman et al., 2004; McCullough, 2002; Sanyal et al., 2004; Vajro et al., 2004) used antioxidants for steatohepatitis.

A preliminary report of a controlled trial that compared vitamin E alone with vitamin E plus pioglitazone said that aminotransferases in patients treated with vitamin E were lower.

Histological improvement, however, was seen only with combined therapy (Sanyal et al., 2004). Patients were randomly treated either with vitamins E and C (1000 IU and 1000 mg, respectively) or with placebo daily for six months. Vitamin treatment resulted in a statistically significant improvement in the fibrosis score without significant side effects.

Drugs That Counter Insulin Resistance

Metformin was associated with a significantly higher normalization of serum ALT versus vitamin E, and both ultrasonography and liver biopsies showed improvements in liver histology and a reduction in fatty infiltration.

Pioglitazone was associated with significant declines in serum aminotransferase levels and increased hepatic insulin sensitivity (Angelico et al., 2007; Belfort et al., 2006; Bugianesi et al., 2005; Uygun et al., 2004).

Probucol

Probucol, a lipid-lowering agent with strong antioxidant properties, significantly reduced ALT levels in patients with NASH (Merat et al., 2003).

Betaine

Betaine (also called *betaine anhydrous* and *trimethylglycine* [TMG]), is a normal component of the metabolic cycle of methionine, which protects against steatosis in animal models. Patients with NASH were treated with an oral solution of betaine, which resulted in a significant normalization in the serum levels of aspartate aminotransferase.

The degree of steatosis, grade of necroinflammation, and stage of fibrosis all fell significantly (Bugianesi et al., 2005).

Twelve months of betaine treatment in 10 adults was associated with nonsignificant attenuation of steatosis.

Ursodeoxycholic Acid

Treating NASH with 13 to 15 mg/kg/d of ursodeoxycholic acid (UDCA) and hypertriglyceridemia with 2 g/d of clofibrate for 12 months, serum alkaline phosphatase, ALT, and gamma-glutamyl transpeptidase levels, as well as the histological grade of steatosis, fell significantly (Laurin et al., 1996).

Probiotics

Probiotics have been proposed as a treatment option for patients with NAFLD and NASH because they counteract the flora in the gut that is a potential source of hepatotoxic oxidative injury (Solga and Diehl, 2003).

Probiotics may be well-tolerated, may improve conventional liver function tests, and may decrease markers of lipid peroxidation (Loguercio et al., 2005, 2007).

Angiotensin II Receptor Blockers

Angiotensin II is involved in the pathogenesis of hepatic fibrosis and increases iron deposition and insulin resistance (Andersen et al., 1991). Forty-eight weeks of treatment with losartan (50 mg/d), an angiotensin II receptor blocker, reduced blood marker levels of hepatic fibrosis, of plasma TGF-$\beta 1$, and of serum ferritin and serum aminotransferase.

Histopathological analysis revealed reduced hepatic necrosis, inflammation, and fibrosis; in addition, it showed no iron deposition.

Losartan treatment had no side effects during the course of the study. Thus, losartan may be therapeutically effective against NASH (Yokohama et al., 2004).

Orlistat

Orlistat is a gastrointestinal lipase inhibitor used to treat obesity and type 2 DM. Six months of orlistat treatment (120 mg tid) reversed fatty infiltration and attenuated hepatic fibrosis and inflammation in 14 obese patients with NASH (Hussein et al., 2007).

The same treatment in another study (Zelber-Sagi et al., 2006) raised serum glucose and insulin in patients with a higher degree of hepatic fibrosis.

However, serum ALT levels and ultrasound showed a reversal of fatty liver in 52 patients with NAFLD.

Sesame Oil

Description

Sesame oil, extracted from the seeds of *Sesamum indicum* (Pedaliaceae family), is a nutrient-rich antioxidant popular in traditional and alternative medicine. It contains sesamin, sesamol, and sesamolin, all of which contribute to its antioxidant property (White, 1992). Sesame has been used for millennia in Chinese and Indian herbal medicine. Although often recommended as a laxative, sesame was used as early as the 4th century A.D. as a Chinese folk remedy for toothache and gum disease. All traditional and alternative medicine considers sesame oil a dietary supplement, nutraceutical (functional food), pharmaceutical aid, and a base or adjuvant.

Sesame oil's medicinal applications are referred to in the traditional medical texts of India and China. In Ayurveda, Indian traditional medicine, a large number of health rejuvenating formulations like Rasayana and Chyavanaprasam, and massage oils in the Thailam class contain sesame oil as the major ingredient (90%) and oil base (Sukumar et al., 2008).

Beneficial Effects of Sesame Oil

Sesame oil is effective against various diseases, e.g., atherosclerosis and hypertension, and against the effects of aging (Namiki et al., 1995). It is also a better antioxidant than canola oil (Baba et al., 1998); more protective than are dietary oils extracted from nuts and sunflower seeds against hypertension, hyperlipidemia, and lipid peroxidation because it increases enzymatic and non-enzymatic antioxidants (Sankar et al., 2005); and, combined with its most potent active ingredient sesamol, is a stronger antitumor agent than are resveratrol and sunflower seed oil (Kapadia et al., 2002). Like sesamol, sesame oil is immediately absorbed by the gastrointestinal system (Periasamy et al., 2013), starting from the oral cavity, which is evident through oil pulling (Asokan et al., 2011). Experiments with rats and mice in our laboratory have shown that sesame oil is protective as well as therapeutically effective within 3-12 h after a sesame-oil gavage (Chandrasekaran et al., 2008; Hsu et al., 2006, 2008, 2009). Sesame oil contains sesamin, sesamol, and sesamolin, all of which are active phenolics known to be hepato-protective. Sesamin attenuates hepatic ischemic-reperfusion injury by inducing both antioxidative and anti-inflammatory activities (Utsunomiya et al., 2003), and it prevents lipid accumulation in the liver (Akimoto et al., 1993). Sesamol protects against acetaminophen-induced liver damage by maintaining glutathione levels and inhibiting lipid peroxidation (Chandrasekaran et al., 2011), and attenuates sinusoidal obstructive syndrome (SOS) by inhibiting MMP-2 and MMP-9 expression, and increasing TIMP-1 expression (Periasamy et al., 2011a, b). Sesamolin reduces serum and liver lipid levels and increases hepatic fatty acid oxidation (Lim et al., 2011).

SESAME OIL ON THE FIRST AND SECOND HITS OF NASH

Sesame oil attenuates methionine-choline deficient (MCD) diet-induced steatohepatitis in mice. Sesame oil decreases steatosis and liver injury, and increased peroxisome proliferator-activated receptor (PPAR)-α, which characterizes the attenuation of the first hit in NASH. Treatment with sesame oil significantly decreases nitric oxide, malondialdehyde (MDA), tumor necrosis factor (TNF)-α, interleukin (IL)-6, IL-1β, leptin, and tissue growth factor (TGF)-β1, which indicates the attenuation of the second hit. Therefore, sesame oil attenuates the parameters involved in the first and second hits of NASH. Sesame oil might intervene in the sequence of pathogenic events of MCD-induced NASH by attenuating early lipid accumulation and peroxidation. Sesame oil decreases leptin expression that might increase stearoyl-CoA desaturase-1 expression, which adds more double bonds to saturated fatty acids and channels them to oxidation, thereby reducing steatosis, lipid synthesis, and storage. Sesame oil's effect on leptin might also decrease TGF-β1 and TNF-α, and reduce inflammation and the development of fibrosis. The decrease in TGF-β1 expression might decrease collagen synthesis and the overproduction of activated hepatic stellate cells, thereby attenuating fibrosis (Periasamy et al., 2014a).

Therapeutic sesame oil attenuates fibrosing steatohepatitis in mice, and it reverses fibrosis by increasing peroxisome proliferator-activated receptor (PPAR)-γ and decreasing collagen deposition.

Liver fibrosis is a dynamic wound-healing process characterized by the increased production and deposition of ECM, mainly collagen and α-smooth muscle actin (α-SMA). Sesame oil increases the expression of PPAR-γ, and it might inhibit TGF-β1 signaling, collagen deposition, and α-SMA expression, which will subsequently reverse fibrosis.

Sesame oil inhibits matrix metalloproteinase-2 and -9, increases the expression of TIMP-1, and attenuates NASH-induced fibrosis. It maintains hepatic tissue homeostasis by controlled and limited proteolytic degradation of the ECM. The dynamics of the ECM in hepatic tissue are balanced between matrix breakdown, which is mediated by MMP, and matrix protein synthesis, which mainly involves collagen deposition and α-SMA. Sesame oil is crucial for balancing ECM degradation and synthesis by activating proteinases and their inhibitors. The inhibition of MMP-2 and -9 expression and upregulation of TIMP-1 expression reduces hepatic injury, hepatic stellate cell activation, ECM degradation, and collagen deposition.

This heals hepatic injury by decreasing AST and ALT activity, which ultimately attenuates fibrosis (Periasamy et al., 2014b).

Sesame oil has the potential to mitigate NASH in humans; however, the US Food and Drug Administration have not approved any medications for treating NASH (Chalasani et al., 2012). The standard of care for treating NASH patients focuses on lifestyle interventions, particularly diet and exercise (Malinowski et al., 2013). The best success is seen with an integrative approach, personalized for individual patients (Afdhal et al., 2012). Therapies for NASH include antioxidants, cytoprotective agents, and insulin sensitizers; however, studies are limited, and these therapies produce adverse effects (Afdhal et al., 2012).

Sesame oil is a non-toxic nutritional oil and effective against various disease models, and it protects against multiorgan failure (Hsu et al., 2002).

Sesame oil therefore can be used clinically to treat NASH in patients undergoing one or more of the current synthetic therapies (Periasamy et al., 2014a, b).

References

Abdelmalek, M. F., Angulo, P., Jorgensen, R. A., Sylvestre, P. B., Lindor, K. D. Betaine, a promising new agent for patients with nonalcoholic steatohepatitis: results of a pilot study. *The American Journal of Gastroenterology,* 2001 96, 2711-2717.

Abdelmalek, M. F., Sanderson, S. O., Angulo, P., Soldevila-Pico, C., Liu, C., Peter, J., Keach, J., Cave, M., Chen, T., McClain, C. J., Lindor, K. D. Betaine for nonalcoholic fatty liver disease: results of a randomized placebo-controlled trial. *Hepatology*, 2009 50, 1818-1826.

Afdhal, N. H. Management of nonalcoholic fatty liver disease: a 60-year-old man with probable nonalcoholic fatty liver disease: weight reduction, liver biopsy, or both? *The Journal of the American Medical Association*, 2012 308, 608-616.

Akimoto, K., Kitagawa, Y., Akamatsu, T., Hirose, N., Sugano, M., Shimizu, S., Yamada, H. Protective effects of sesamin against liver damage caused by alcohol or carbon tetrachloride in rodents. *Annals of Nutrition and Metabolism,* 1993 37, 218-224.

Andersen, T., Gluud, C., Franzmann, M. B., Christoffersen, P. Hepatic effects of dietary weight loss in morbidly obese subjects. *Journal of Hepatology*, 1991 12, 224-229.

Angelico, F., Burattin, M., Alessandri, C., Del Ben, M., Lirussi, F. Drugs improving insulin resistance for non-alcoholic fatty liver disease and/or non-alcoholic steatohepatitis. *Cochrane Database of Systematic Reviews*, 2007 1, CD005166.

Angulo, P., Lindor, K. D. Non-alcoholic fatty liver disease. *Journal of Gastroenterology and Hepatology,* 2002 17(Suppl.), S186-S190.

Asokan, S., Kumar, R. S., Emmadi, P., Raghuraman, R., Sivakumar, N. Effect of oil pulling on halitosis and microorganisms causing halitosis: a randomized controlled pilot trial. *Journal of Indian Society of Pedodontics and Preventive Dentistry*, 2011 29, 90-94.

Baba, N. H., Antoniades, K., Habbal, Z. Effects of dietary canola, olive, and linolenic acid enriched olive oils on plasma lipids, lipid peroxidation and lipoprotein lipase activity in rats. *Nutrition Research* 1998 49, 41-45.

Belfort, R., Harrison, S. A., Brown, K., Darland, C., Finch, J., Hardies, J., Balas, B., Gastaldelli, A., Tio, F., Pulcini, J., Berria, R., Ma, J. Z., Dwivedi, S., Havranek, R., Fincke, C., DeFronzo, R., Bannayan, G. A., Schenker, S., Cusi, K. A placebo-controlled trial of pioglitazone in subjects with nonalcoholic steatohepatitis. *The New England Journal of Medicine* 2006 355, 2297-2307.

Bellentani, S., Scaglioni, F., Marino, M., Bedogni, G. Epidemiology of non-alcoholic fatty liver disease. *Digestive Diseases and Sciences* 2010 28, 155-161.

Bugianesi, E., Gentilcore, E., Manini, R., Natale, S., Vanni, E., Villanova, N., David, E., Rizzetto, M., Marchesini, G. A randomized controlled trial of metformin versus vitamin E or prescriptive diet in nonalcoholic fatty liver disease. *The American Journal of Gastroenterology*, 2005 100, 1082-1090.

Chalasani, N., Younossi, Z., Lavine, J. E., Diehl, A. M., Brunt, E. M., Cusi, K., Charlton, M., Sanyal, A. J. The diagnosis and management of nonalcoholic fatty liver disease: practice Guideline by the American Association for the Study of Liver Diseases, American College of Gastroenterology, and the American Gastroenterological Association. *Hepatology*, 2012 55, 2005-2023.

Chandrasekaran, V. R., Chien, S. P., Hsu, D. Z., Liu, M. Y. Anti-hepatotoxic effects of 3, 4-methylenedioxyphenol and N-acetylcysteine in acutely acetaminophen-overdosed mice. *Human and Experimental Toxicology*, 2011 30, 1609-1615.

Chandrasekaran, V. R., Wan, C. H., Liu, L. L., Hsu, D. Z., Liu, M. Y. Effect of sesame oil against acetaminophen-induced acute oxidative hepatic damage in rats. *Shock* 2008 30, 217-221.

Chitturi, S., Farrell, G. C. Etiopathogenesis of nonalcoholic steatohepatitis. *Seminars in Liver Disease*, 2001 21, 27-41.

Cusi, K. Role of insulin resistance and lipotoxicity in non-alcoholic steatohepatitis. *Clinics in Liver Disease*, 2009 13, 545-563.

Dansinger, M. L., Gleason, J. A., Griffith, J. L., Selker, H. P., Schaefer, E. J. Comparison of the Atkins, Ornish, Weight Watchers, and Zone diets for weight loss and heart disease risk reduction: a randomized trial. *The Journal of the American Medical Association,* 2005 293, 43-53.

Falchuk, K. R., Fiske, S. C., Haggitt, R. C., Federman, M., Trey, C. Pericentral hepatic fibrosis and intracellular hyalin in diabetes mellitus. *Gastroenterology*, 1980 78, 535-541.

George, J., Pera, N., Phung, N., Leclercq, I., Yun Hou, J., Farrell, G. Lipid peroxidation, stellate cell activation and hepatic fibrogenesis in a rat model of chronic steatohepatitis. *Journal of Hepatology*, 2003 39, 756-764.

Hickman, I. J., Jonsson, J. R., Prins, J. B., Ash, S., Purdie, D. M., Clouston, A. D., Powell, E. E. Modest weight loss and physical activity in overweight patients with chronic liver disease results in sustained improvements in alanine aminotransferase, fasting insulin, and quality of life. *Gut*, 2004 53, 413-419.

Hornboll, P., Olsen, T. S. Fatty changes in the liver: the relation to age, overweight and diabetes mellitus. *Acta Pathologica et Microbiologica Scandinavica. Section A*, 1982 90, 199-205.

Hsu, D. Z., Chen, K. T., Chien, S. P., Li, Y. H., Huang, B. M., Chuang, Y. C., Liu, M. Y. Sesame oil attenuates acute iron-induced lipid peroxidation-associated hepatic damage in mice. *Shock*, 2006 26, 625-630.

Hsu, D. Z., Chien, S. P., Li, Y. H., Chuang, Y. C., Chang, Y. C., Liu, M. Y. Sesame oil attenuates hepatic lipid peroxidation by inhibiting nitric oxide and superoxide anion generation in septic rats. *JPEN Journal of Parenteral and Enteral Nutrition,* 2008 32, 154-159.

Hsu, D. Z., Chu, P. Y., Liu, M. Y. Effect of sesame oil on acidified ethanol-induced gastric mucosal injury in rats. *JPEN Journal of Parenteral and Enteral Nutrition*, 2009 33, 423-427.

Hsu, D. Z., Liu, M. Y. Sesame oil attenuates multiple organ failure and increases survival rate during endotoxemia in rats. *Critical Care Medicine*, 2002 30, 1859-1862.

Huang, M. A., Greenson, J. K., Chao, C., Anderson, L., Peterman, D., Jacobson, J., Emick, D., Lok, A. S., Conjeevaram, H. S. One-year intense nutritional counseling results in histological improvement in patients with nonalcoholic steatohepatitis: a pilot study. *The American Journal of Gastroenterology,* 2005 100, 1072-1081.

Hussein, O., Grosovski, M., Schlesinger, S., Szvalb, S., Assy, N. Orlistat reverses fatty infiltration and improves hepatic fibrosis in obese patients with nonalcoholic steatohepatitis (NASH). *Digestive Diseases and Sciences,* 2007 52, 2512-2519.

Kapadia, G. J., Azuine, M. A., Tokuda, H., Takasaki, M., Mukainaka, T., Konoshima, T., Nishino, H. Chemopreventive effect of resveratrol, sesamol, sesame oil and sunflower oil in the Epstein-Barr virus early antigen activation assay and the mouse skin two-stage carcinogenesis. *Pharmacological Research*, 2002 45, 499-504.

LaBrecque, D. R., Abbas, Z., Anania, F., Ferenci, P., Khan, A. G., Goh, K.-L., Hamid, S. S., Isakov, V., Lizarzabal, M., Peñaranda, M. M., Ramos, J. F. R., Sarin, S., Stimac, D., Thomson, A. B. R., Umar, M., Krabshuis, J., LeMair, A., Review Team. World Gastroenterology Organisation Global Guidelines: nonalcoholic fatty liver disease and nonalcoholic steatohepatitis. *Journal of Clinical Gastroenterology*, 2014 48, 467-473.

Laurin, J., Lindor, K. D., Crippin, J. S., Gossard, A., Gores, G. J., Ludwig, J., Rakela, J., McGill, D. B. Ursodeoxycholic acid or clofibrate in the treatment of non-alcohol-induced steatohepatitis: a pilot study. *Hepatology*, 1996 23, 1464-1467.

Lim, J. S., Adahi, Y., Takahashi, Y., Ide, T. Comparative analysis of sesame lignans (sesamin and sesamolin) in affecting hepatic fatty acid metabolism in rats. *British Journal of Nutrition*, 2007 97, 85-95.

Lirussi, F., Mastropasqua, E., Orando, S., Orlando, R. Probiotics for non-alcoholic fatty liver disease and/or steatohepatitis. *Cochrane Database of Systematic Reviews,* 2007 1, CD005165.

Loguercio, C., Federico, A., Trappoliere, M., Tuccillo, C., de Sio, I., Di Leva, A., Niosi, M., D'Auria, M. V., Capasso, R., Del Vecchio Blanco, C., Real Sud Group. The effect of a silybin-vitamin E-phospholipid complex on nonalcoholic fatty liver disease: a pilot study. *Digestive Diseases and Sciences*, 2007 52, 2387-2395.

Loguercio, C., Federico, A., Tuccillo, C., Terracciano, F., D'Auria, M. V., De Simone, C., Del Vecchio Blanco, C. Beneficial effects of a probiotic VSL#3 on parameters of liver dysfunction in chronic liver diseases. *Journal of Clinical Gastroenterology*, 2005 39, 540-543.

Ludwig, J., Viggiano, T. R., McGill, D. B., Oh, B. J. Nonalcoholic steatohepatitis: Mayo Clinic experiences with a hitherto unnamed disease. *Mayo Clinic Proceedings*, 1980 55, 434-438.

MacDonald, G. A., Bridle, K. R., Ward, P. J., Walker, N. I., Houglum, K., George, D. K., Smith, J. L., Powell, L. W., Crawford, D. H., Ramm, G. A. Lipid peroxidation in hepatic steatosis in humans is associated with hepatic fibrosis and occurs predominately in acinar zone 3. *Journal of Gastroenterology and Hepatology,* 2001 16, 599-606.

Malinowski, S. S., Byrd, J. S., Bell, A. M., Wofford, M. R., Riche, D. M. Pharmacologic therapy for nonalcoholic fatty liver disease in adults. *Pharmacotherapy*, 2013 33, 223-242.

McCullough, A. J. Update on nonalcoholic fatty liver disease. *Journal of Clinical Gastroenterology,* 2002 34, 255-262.

Merat, S., Malekzadeh, R., Sohrabi, M. R., Hormazdi, M., Naserimoghadam, S., Mikaeli, J., Farahvash, M. J., Ansari, R., Sotoudehmanesh, R., Khatibian, M. Probucol in the treatment of nonalcoholic steatohepatitis. *Journal of Clinical Gastroenterology*, 2003 36, 266-268.

Murray, M., Zaluzny, L., Farrell, G. C. Drug metabolism in cirrhosis. Selective changes in cytochrome P-450 isozymes in the choline deficient rat model. *Biochemical Pharmacology*, 1986 35, 1817-1824.

Namiki, M. The chemistry and physiological function of sesame. *Food Research International,* 1995 11; 281-329.

Naniwadekar, A. S. Nutritional recommendations for patients with non-alcoholic fatty liver disease: An evidence based review. *Gastroenterology Research and Practice*, 2010, 8-16.

Neuschwander-Tetri, B. A., Caldwell, S. H. Nonalcoholic steatohepatitis: summary of an AASLD single topic conference. *Hepatology*, 2003 37, 1202-1219.

Nobili, V., Manco, M., Devito, R., Di Ciommo, V., Comparcola, D., Sartorelli, M. R., Piemonte, F., Marcellini, M., Angulo, P. Lifestyle intervention and antioxidant therapy in children with nonalcoholic fatty liver disease: a randomized, controlled trial. *Hepatology*, 2008 48, 119-128.

Periasamy, S., Chien, S. P., Chang, P. C., Hsu, D. Z., Liu, M. Y. Sesame oil mitigates nutritional steatohepatitis via attenuation of oxidative stress and inflammation: a tale of two-hit hypothesis. *The Journal of Nutritional Biochemistry*, 2014a 25, 232-240.

Periasamy, S., Hsu, D. Z., Chang, P. C., Liu, M. Y. Sesame oil attenuates nutritional fibrosing steatohepatitis by modulating matrix metalloproteinases-2, 9 and PPAR-γ. *The Journal of Nutritional Biochemistry,* 2014b 25, 337-344.

Periasamy, S., Hsu, D. Z., Chen, S. Y., Yang, S. S., Chandrasekaran, V. R., Liu, M. Y. Therapeutic sesamol attenuates monocrotaline-induced sinusoidal obstruction syndrome in rats by inhibiting matrix metalloproteinase-9. *Cell Biochemistry and Biophysics*, 2011a 61, 327-336.

Periasamy, S., Mo, F. E., Chen, S. Y., Chang, C. C., Liu, M. Y. Sesamol attenuates isoproterenol induced acute myocardial infarction via inhibition of matrix metalloproteinase-2 and -9 expression in rats. *Cellular Physiology and Biochemistry,* 2011b 27, 273-280.

Periasamy, S., Yang, S. S., Chen, S. Y., Chang, C. C., Liu, M. Y. Prophylactic sesame oil attenuates sinusoidal obstruction syndrome by inhibiting matrix metalloproteinase-9 and oxidative stress. *JPEN Journal of Parenteral and Enteral Nutrition,* 2013 37, 529-537.

Peters, R. L., Gay, T., Reynolds, T. B. Post-jejunoileal bypass hepatic disease. Its similarity to alcoholic liver disease. *American Journal of Clinical Pathology,* 1975 63, 318-331.

Rinella, M. E., Green, R. M. The methionine-choline deficient dietary model of steatohepatitis does not exhibit insulin resistance. *Journal of Hepatology,* 2004 40, 47-51.

Sankar, D., Sambandam, G., Ramakrishna Rao, M., Pugalendi, K. V. Modulation of blood pressure, lipid profiles and redox status in hypertensive patients taking different edible oils. *Clinica Chimica Acta*, 2005 355, 97-104.

Sanyal, A., Mofrad, P. S., Contos, M. J., Sargeant, C., Luketic, V. A., Sterling, R. K., Stravitz, R. T., Shiffman, M. L., Clore, J., Mills, A. S. A pilot study of vitamin E versus vitamin E and pioglitazone for the treatment of nonalcoholic steato-hepatitis. *Clinical Gastroenterology and Hepatology*, 2004 2, 1107-1115.

Sanyal, A. J. American Gastroenterological Association. AGA technical review on nonalcoholic fatty liver disease. *Gastroenterology*, 2002 123, 1705-1725.

Seki, S., Kitada, T., Yamada, T., Sakaguchi, H., Nakaatani, K., Wakasa, K. In situ detection of lipid peroxidation and oxidative DNA damage in non-alcoholic fatty liver diseases. *Journal of Hepatology*, 2002 37, 56-62.

Solga, S. F., Diehl, A. M. Non-alcoholic fatty liver disease: lumen-liver interactions and possible role for probiotics. *Journal of Hepatology*, 2003 38, 681-687.

Sukumar, D., Arimboor, R., Arumughan, C. HPTLC fingerprinting and quantification of lignans as markers in sesame oil and its polyherbal formulations. *Journal of Pharmaceutical and Biomedical Analysis*, 2008 47, 795-801.

Tilg, H., Diehl, A. M. Cytokines in alcoholic and nonalcoholic steatohepatitis. *The New England Journal of Medicine*, 2000 343, 1467-1476.

Utsunomiya, T., Shimada, M., Rikimaru, T., Hasegawa, H., Yamashita, Y., Hamatsu, T., Yamasaki, M., Kaku, S., Yamada, K., Sugimachi, K. Antioxidant and anti-inflammatory effects of a diet supplemented with sesamin on hepatic ischemia-reperfusion injury in rats. *Hepatogastroenterology*, 2003 50, 1609-1613.

Uygun, A., Kadayifci, A., Isik, A. T., Ozgurtas, T., Deveci, S., Tuzun, A., Yesilova, Z., Gulsen, M., Dagalp, K. Metformin in the treatment of patients with non-alcoholic steatohepatitis. *Alimentary Pharmacology and Therapeutics*, 2004 19, 537-544.

Vajro, P., Mandato, C., Franzese, A., Ciccimarra, E., Lucariello, S., Savoia, M., Capuano, G., Migliaro, F. Vitamin E treatment in pediatric obesity-related liver disease: a randomized study. *Journal of Pediatric Gastroenterology and Nutrition,* 2004 38, 48-55.

White, J. P. Fatty acids in oil seeds (vegetable oils). In: C. K. Chow (eds.) *Fatty Acids in Foods and Their Health Implications*. New York: M. Dekker, 1992, pp. 237-262.

Yokohama, S., Yoneda, M., Haneda, M., Okamoto, S., Okada, M., Aso, K., Hasegawa, T., Tokusashi, Y., Miyokawa, N., Nakamura, K. Therapeutic efficacy of an angiotensin II receptor antagonist in patients with nonalcoholic steatohepatitis. *Hepatology*, 2004 40, 1222-1225.

Zelber-Sagi, S., Kessler, A., Brazowsky, E., Webb, M., Lurie, Y., Santo, M., Leshno, M., Blendis, L., Halpern, Z., Oren, R. A double-blind randomized placebo-controlled trial of orlistat for the treatment of nonalcoholic fatty liver disease. *Clinical Gastroenterology and Hepatology*, 2006 4, 639-644.

INDEX

A

B

C

D

E

M

Q

R

S

T

R

S

T

U

V

W

X

Y

α